INTEGRATED SYSTEM-LEVEL MODELING OF
NETWORK-ON-CHIP ENABLED MULTI-PROCESSOR PLATFORMS

Integrated System-Level Modeling of Network-on-Chip enabled Multi-Processor Platforms

Tim Kogel

CoWare, Aachen, Germany

Rainer Leupers

RWTH Aachen, Germany

Heinrich Meyr

RWTH Aachen, Germany

A C.I.P. Catalogue record for this book is available from the Library of Congress.

ISBN-13 978-90-481-7202-3
ISBN-10 1-4020-4826-2 (e-books)
ISBN-13 978-1-4020-4826-2 (e-books)

Published by Springer,
P.O. Box 17, 3300 AA Dordrecht, The Netherlands.

www.springer.com

Printed on acid-free paper

Printed in the Netherlands.

Gewidmet meiner Frau Miriam,
meinen Söhnen Leon und Nathan, und
meinen Eltern Walter und Renate.

Contents

Foreword

We are presently observing a paradigm change in designing complex SoC as it occurs roughly every twelve years due to the exponentially increasing number of transistors on a chip. This design discontinuity, as all previous ones, is characterized by a move to a higher level of abstraction. This is required to cope with the rapidly increasing design costs. While the present paradigm change shares the move to a higher level of abstraction with all previous ones, there exists also a key difference. For the first time shrinking geometries do not lead to a corresponding increase of performance. In a recent talk Lisa Su of IBM pointed out that in 65nm technology only about 25% of performance increase can be attributed to scaling geometries while the lion share is due to innovative processor architecture [1]. We believe that this fact will revolutionize the entire semiconductor industry.

What is the reason for the end of the traditional view of Moore's law? It is instructive to look at the major drivers of the semiconductor industry: wireless communications and multimedia. Both areas are characterized by a rapidly increasing demand of computational power in order to process the sophisticated algorithms necessary to optimally utilize the precious resource bandwidth. The computational power cannot be provided by traditional processor architectures and shared bus type of interconnects. The simple reason for this fact is energy efficiency: there exist orders of magnitude between the energy efficiency of an algorithm implemented as a fixed functionality computational element and of a software implementation on a processor.

We argue that future SoC for wireless and multimedia applications will be implemented as heterogeneous multiprocessor systems (MP-SoC) in order to achieve an optimum in the trade-off between energy efficiency versus flexibility (programmability). Such an optimum trade-off is ultimately necessary to cope with the required flexibility of multi-standard, cognitive software defined radio which promotes a software implementation. The heterogeneous MP-SoC will contain an increasing number of application specific processors

(ASIPs) combined with complex memory hierarchies and sophisticated on chip communication networks.

The design of an MP-SoC is an extremely demanding task. Already in 2001 ITRS has pointed out that "The main message in 2001 is this: Cost of design is the greatest threat to continuation of the semiconductor roadmap". In a nutshell, designing an MP-SoC comprises two major tasks. The first task is to define a set of processing elements which perform the energy efficient execution of the functional task. The second, and equally important, task is concerned with the inter-task data exchanges which have to be mapped onto an interconnect architecture. Both computation and communication have seen significant advances in terms of functionality and architectural concepts. As a result, also the mapping of an application onto a MP-SoC platform becomes an increasingly demanding task. Only a joint consideration of architectural options and application mapping bears the opportunity to achieve near optimal quality of results.

In this book we have made an attempt to present a unified system level design framework for the definition and programming of large scale, heterogeneous MP-SoC platforms. This comprises the exploration of architectural choices for computation and communication as well as for the HW/SW partitioning and mapping of embedded applications. One focus area is the emerging topic of Network-on-Chips, which are envisioned to become the communication backbone of next generation Multi-Processor platforms.

The huge literature on the subject is scattered in journals and conference publications and thus not readily accessible to the engineer in industry. We therefore first give a fairly broad introduction to classify the topic in terms of application domains, architectural elements and system level design methods. We hope by this to provide the reader with a reasonably efficient path towards gaining an understanding of the subject. We have also made an attempt to cover the state of the art research results by including the most recent publications. We hope that this book will be useful to the engineer in industry who wants to get an overview of the latest trends in SoC architectures and system-level design methodologies. We also hope that this book will be useful to academia actively engaged in research.

HEINRICH MEYR AND RAINER LEUPERS, FEBRUARY 2006

Preface

This book documents more than 5 years of research during my time as a research assistant at the Institute for Integrated Signal Processing Systems (ISS) at the Aachen University of Technology (RWTH Aachen).

The original motivation for this work dates back to the middle 1990ies. It was driven by the attempt to define an holistic approach to the design of algorithms, tools, and architectures for an Asynchronous Transfer Mode (ATM) backbone packet switch. At that time, system level design methodologies were still in their infancy, but the complexity to design this type of heterogeneous Hardware/Software systems was already getting out of control.

When I joined the team in 1999, the early work on the ATM packet switch had already created a wealth of experience on abstract C-based modeling of complex architectures. Building on this know-how, we soon ported our research results to the newly available SystemC library. The move to a standardized modeling language enabled a number of further research cooperations with different industrial partners. During these projects we have evolved our design methodology and tools as well as broadened the application domain beyond the original networking space. Even more importantly, we were able to validate our approach in the context of real-life industrial design problems.

Looking back, the results presented in this book are by no means attributed to some stroke of brilliance or the like of it, but rather the evolutionary development of many small steps towards mastering the SoC complexity crisis. In the following I would like to thank the many brilliant and open-minded people from the ISS institute and our industrial research partners, with whom I had the pleasure to work and who have made invaluable contributions to the content of this book, be it through focus and advise or actual hands-on work.

At the outset I would like to thank Prof. Heinrich Meyr as the supervisor of my research activities. Besides his ongoing personal interest in my work, he has created an atmosphere of competition and support, which in combination with a tight industrial interaction enables both relevant and state-of-the-art research results. In the same way I like to thank Prof. Rainer Leupers and Prof. Gerd Ascheid, who joined the ISS and gave me the same type of support. I am also thankful to Prof. Perti Mähönen for the valuable feedback he gave me in his role as the additional supervisor of my thesis.

The ground-work for the results described in this book was done by my predecessors Dr. Guido Post and Dr. Andrea Kroll. Apart from providing an excellent starting point, my special thanks is directed to Andrea, who supervised my master thesis in 1998, afterwords recruited me to the ISS and was my mentor during my first two years as a research assistant at the institute.

A major share of the effort to turn the concepts described in this book into actual tangible results is attributed to the master students, who contributed with their skills and their hard work. For their personal engagement I like to thank (in alphabetical order) Malte Dörper, Torsten Kempf, Roland Nennen, Thomas Philipp, Andreas Wieferink, and Olaf Zerres.

I personally consider the ongoing deployment of the tools and methodologies in the context of industrial cooperations as the major advantage for validating the relevance and applicability of any engineering research. During these projects I received invaluable feedback and guidance from a large number of professionals throughout the semiconductor and EDA industries. Among these I especially like to thank Bernd Reinkemeier, Dr. Thorsten Grötker, and Dr. Martin Vaupel from Synopsys, Hans-Jürgen Reumermann from Philips, as well as Kakimotosan, Tangi-san, and Tsunakava-san from Sony.

I was fortunate to be able to continue the work on this topic during my subsequent life at CoWare Inc. Here the concepts and prototype tools described in this book have been turned into a commercial product. The resulting *Architects View Framework* is now available as an option of the CoWare Platform Architect product. I like to thank all the people in CoWare, who have contributed to this effort, including Pascal Chauvet, Malte Dörper, Dr. Serge Goossens, Eshel Haritan, Aldwin Keppens, Igor Makovicky, Xavier Van Elsacker, Dr. Karl VanRompaey, and Bart Vanthournout.

I am especially grateful for the refuge from daily's stressful life my parents provided during the period of writing all this down. Most importantly I like to thank my wife for her constant support, confidence, and love.

TIM KOGEL, FEBRUARY 2006

Chapter 1

INTRODUCTION

Traditionally, embedded applications in the multimedia, wireless communications or networking domain have been implemented on Printed Circuit Boards (PCBs). PCB systems are composed of discrete Integrated Circuits (ICs) like General Purpose Processors, Digital Signal Processors, Application Specific Integrated Circuits, memories, and further peripherals. The communication between the discrete processing elements and memories is realized by shared bus architectures.

The ongoing progress in silicon technology fosters the transition from board-level integration towards System-on-Chip (SoC) implementations of embedded applications. According to the International Technology Roadmap for Semiconductors [2], by the end of the decade SoCs will grow to 4 billion transistors running at 10 GHz and operating below one volt. Already today multiple heterogeneous processing elements and memories can be integrated on a single chip to increase performance and to reduce cost and improve energy efficiency [3].

The growing potential for silicon integration is even outpaced by the amount of functionality incorporated into embedded devices from all kinds of application domains. This trend originates from the tremendous increase in features as well as the multitude of co-existing standards. The resulting functional complexity clearly promotes Software enabled solutions to achieve the required flexibility and cope with the demanding time-to-market conditions. However, the stringent energy efficiency constraints of mobile applications and cost sensitive consumer devices prohibit the use of general purpose processors. Instead, the tight cost and performance requirements of versatile embedded systems lead to application specific heterogeneous multi-processor architectures [4, 5].

In this context, the classical vertical partitioning approach to HW/SW Co-design, where the performance critical parts are implemented as dedicated HW blocks and the rest is executed in SW, is no longer applicable [6]. Instead HW/SW Co-design can be seen as a multi-dimensional horizontal mapping problem of an application running on a heterogeneous multiprocessor platform.

During the mapping process, the system architect has to exploit application inherent parallelism to achieve the required performance at reasonable cost. For the computationally intensive portions of typical embedded applications the *extraction* of Task Level Parallelism (TLP) is mostly straight forward: The partitioning into a set of loosely coupled functional blocks can be naturally derived from the algorithmic block diagram.

Still the spatial and temporal application-to-architecture mapping poses an enormous challenge in the design of embedded systems. First, a set of processing elements has to be provided for the efficient execution of the functional tasks. Additionally, the inter-task data exchange has to be mapped to a communication architecture. Both processing and communication mapping are highly interrelated and only a joint consideration of architectural choices in both areas bears the opportunity for near optimal quality of results. Especially recent architectural advances offer a huge design space with enormous potential for optimization:

Communication Architectures. Today's predominant shared bus paradigm as inherited from the PCB era constitutes the major power and performance bottleneck. In response to this problem, the chip-wide communication is envisioned to be handled by full-scale Network-on-Chip (NoC) architectures [7]. Dedicated on-chip networks enable the use of physically optimized transmission channels to address power, reliability and performance issues [8, 9].

Apart from resolving the physical issues, Network-on-Chip architectures also address the *functional* aspects of on-chip communication. So far, the dynamic priority based arbitration scheme of shared busses creates a mutual dependency between all components connected to the bus. Due to this lack of traffic management capabilities every change in the traffic requirements of the application requires a re-design of the bus architecture. Instead, NoC architectures take advantage of sophisticated networking algorithms to provide elaborated traffic-management capabilities. By that, the ad-hoc communication mapping is replaced with a disciplined allocation of the required communication services and the on-chip network takes care to provide the required resources.

From the system architecture perspective, this separation of the offered communication services from the architectural resources can be considered as a *virtualization* of the actual communication architecture [10]. This virtualization effectively decouples the mapping problem for communication and computation. The price to pay for the physical and functional benefits of NoC based communication is a significant penalty in terms of chip area as well as transfer latency.

Computational Architectures. Concerning the evolution of computational resources, programmable processing elements achieve significant gains with respect to performance and computational efficiency by tailoring instruction set and micro architecture to the respective set of tasks [11]. Examples are innovative architectures exploiting Instruction Level Parallelism (ILP) as well as Data Level Parallelism (DLP) [12]. Despite the increased computational performance, the effective performance is often constricted by the communication architecture, since memory accesses latency does not keep pace with the processing power.

General purpose processors resolve the memory access bottleneck by using sophisticated cache and memory hierarchies. Unfortunately this approach is often not applicable for embedded applications due to the poor memory locality of stream driven and packet based data processing. Instead, processor architectures are equipped with hardware supported Multi-Threading (HW-MT) [13] to perform task switches with virtually no performance overhead. By that, the application inherent TLP is exploited with the purpose of hiding memory latency, which effectively leads to a significant increase in the processor utilization. This technique is already widely employed in the network processor domain [14] but recently finds its way into advanced multimedia [15] and signal processing platforms [16]. In the light of the latency issue caused by NoC architectures, the importance of memory hiding techniques is likely to increase in the future.

Apart from the immediate benefit of increased utilization, HW-MT can be considered as a lean operating system implemented in hardware to efficiently share the processing resources among multiple concurrent tasks. In analogy with full scale software operating systems (SW-OS), the HW-MT concept bears the potential to bring a disciplined management of processing resources to the data processing domain. From the perspective of the functional tasks, this processing management again introduces a virtualization of the computational resources. [17]

Taking the above considerations together, future SoCs can be considered as *NoC enabled multi-processor architectures*. The on-chip communication backbone connects a large number of heterogeneous processing clusters and global storage elements. Individual processing clusters consist of one or few application specific programmable kernels together with tightly coupled instruction and data memories as well as local peripherals.

Design Complexity. The key concept to cope with the resulting design complexity is to achieve a virtualization of the architectural resources, such that they can be *allocated* by the system architect in a deterministic way. As discussed above, this virtualization is provided by the novel NoC approach for the communication part as well as by SW and HW operating systems for the control and data processing respectively. This divide-and-conquer oriented design paradigm enables individual optimization of the architectural elements to take full advantage of recent developments in computer architecture and NoC enabled communication. The price for these benefits with respect to both design efficiency and architectural efficiency is merely a penalty in terms of chip area, which is generally considered to be of constantly decreasing importance.

In this context HW/SW Co-design of a given embedded application is defined to a) architect a heterogeneous MP-SoC platform and b) allocate the architectural resources for the execution of the application. Note, that architecture virtualization resolves the mutual dependencies in the mapping process, but the *trade-offs* in the design space still require a joint consideration of application and architecture as well as communication and communication. For example the latency of a more complex on-chip network can be compensated by either introducing memory hierarchy or employing hardware multi-threaded processor kernels. Obviously, the resulting design space is virtually infinite and the architecting and the mapping phase cannot be considered independently without sacrificing quality of results.

The focus of this book is the introduction of a system level design methodology and corresponding tool supported modeling framework, which together address the multidimensional phase-coupled design space exploration challenge. The goal of this approach is to enable the mapping of the considered application onto the anticipated MP-SoC architectures at a very early stage in the design flow. The modeling framework is based on a sophisticated timing model, which captures the impact on performance of both the computation as well as the communication architecture in a unified and highly abstract way. The achieved accuracy, modeling efficiency and simulation performance enables the exploration of large design spaces, thus the system architect can take full advantage of the architectural innovations outlined above.

The remainder of this section provides a brief overview about the different aspects discussed in this book. First a brief discussion of the abstraction levels clarifies the relation of the proposed approach and the state of the art in System Level Design. Then an intuitive introduction of the timing model is given, which enables an abstract and yet accurate modeling of the anticipated architecture. Later a short introduction illustrates the modular simulation framework for rapid design space exploration of Network-on-Chip enabled heterogeneous MP-SoC platforms.

Abstraction Level. Transaction-Level Modeling (TLM) as advocated by the SystemC language [18] is generally considered as the emerging system level design paradigm and is already incorporated into state-of-the-art Electronic System Level (ESL) tools [19, 20]. TLM greatly improves modeling efficiency and simulation speed by abstracting from the low-level communication details of the Register Transfer Level (RTL), but is usually employed in a byte and cycle accurate fashion.

For the conceptualization of large scale heterogeneous systems as addressed in this book, cycle-level TLM is still too detailed to explore large design spaces. Instead, the developed modeling framework is based on a packet-level TLM paradigm. Here the considered data granularity is a set of functionally associated data items, which are combined into an Abstract Data Type (ADT). This data representation is much closer to the initial application model, so the modeling efficiency as well as the simulation speed are again significantly improved compared to cycle-accurate TLM. The key aspect of this approach is that the underlying timing model outlined below is sufficiently accurate to investigate the performance impact of the anticipated MP-SoC architecture executing the application.

Unified Timing Model. Inspired by the observation, that *communication* becomes the driving design paradigm for MP-SoC from application to architecture mapping [21], the developed exploration framework is based on a sophisticated, communication centric timing model, which can be coarsely separated into the following aspects:

- A generic synchronization interface defines a concise set of communication primitives, which in principle follow the Open Core Open Core Protocol (OCP) semantics [22] and are not biased towards any specific communication architecture. Additionally the primitives incorporate *timing-annotation* to achieve reasonable timing accuracy at the highly abstract packet-level TLM layer.

- The communication timing model captures the impact on performance of the interconnection architecture. This communication timing model supports the full spectrum of available and proposed communication architectures ranging from today's shared busses to the emerging NoC paradigm [23, 24].
- The processing delay annotation virtually maps individual application tasks to the intended processing engines [25]. The resulting impact on performance is captured by calculating the timing of the external events, which are exposed by the generic communication interface.
- The concept of a Virtual Processing Unit (VPU) models the notion of shared coarse-grain computational resources. This covers both software operating systems as well as hardware multi-threading.

Exploration Framework. The unified timing model outlined above is implemented by means of a versatile modeling framework for architecture exploration and hardware/software partitioning. Apart from the modeling efficiency and simulation speed inherent to the high abstraction level, a key aspect for efficient design space exploration is a declarative specification mechanism. By that the various aspects of the MP-SoC platform, like e.g. communication architecture, processing elements and task mapping, are defined by a set of configuration files. As part of the elaboration phase, the developed simulator evaluates the configuration files and constructs the specified architecture. During the simulation run, the simulation framework provides an interactive Graphical User Interface (GUI) based on the Message Sequence Chart (MSC) principle to support the interactive validation of the simulation model. The simulation results like latency, delay and utilization of processing elements and communication links are stored in a data base. This raw data is compiled into a set aggregated histograms and performance graphs by means of statistical post-processing. Based on these results, the system architect can detect bottlenecks or poor utilization in the system and decide on further optimizations of the architecture model.

1.1 Organization of the Book Chapters

The contribution of this work is a unified system level design framework for architectural exploration of large scale, heterogeneous MP-SoC platforms as well as Hardware/Software partitioning of embedded applications. As this topic is extensively addressed by academic research and by EDA companies, first a broad introductory part classifies the topic area in terms of application domains, architectural elements, and system level design methods.

At the outset, a brief overview of major application domains is given in chapter 2 to highlight current and future application requirements. In a similar way, chapter 3 classifies current and emerging MP-SoC architecture components. This comprises processing elements as well as communication architectures. From the discussion of both application and architecture characteristics, the requirements for the design of MP-SoC platforms are derived.

After a brief introduction of fundamentals in system level design like abstraction mechanisms and models of computation in chapter 4, the following chapter 5 surveys the state of the art in the area of system level design methodologies and tooling. This chapter closes with a summarizing discussion of benefits and shortcomings of the related work in academia and industry.

Subsequent to these introductory chapters, the main body of this book is dedicated to the comprehensive description of the contribution. First an intuitive description of the developed MP-SoC framework and associated design methodology is provided in chapter 6. This overview sets the stage for the following chapters containing all the detailed information.

The theoretical foundation of the developed timing model is formulated in chapter 7. After a brief introduction of the employed Tagged Signal Model formalism [26], the timing model is introduced as a derivation of the well-known Discrete Event (DE) Model of Computation (MoC). Afterwords the diverse aspects of timing modeling with respect to communication, computation and multi-threading are covered in detail.

The implementation of the timing model by means of a versatile system level Design Space Exploration (DSE) environment for MP-SoC platforms is described in chapter 8. Major components of this framework are the Network-on-Chip framework for communication modeling and the generic Virtual Processing Unit (VPU) to model multi-threaded processing elements. Additionally, the various visualization mechanisms for functional validation and performance analysis are highlighted.

The applicability of the design space exploration framework and tooling introduced in book is demonstrated by a large scale case-study. The selected IPv4 application with Quality-of-Service (QoS) support as well as key results from the investigation of architectural alternatives are provided in chapter 9.

Finally, chapter 10 summarizes the major achievements of the work described in this book and concludes with an outlook on future developments.

Chapter 2

EMBEDDED SOC APPLICATIONS

Traditionally, applications of embedded systems are classified into different application domains, like networking, multimedia, and wireless communications. This chapter examines applications from different domains in order to derive common properties and requirements with respect to their implementation on MP-SoC platforms. The networking application domain is treated with the highest detail, since the case study elaborated in chapter 9 falls into this category. Additionally, a basic knowledge of networking concepts is helpfull for the understanding of on-chip micro networks.

2.1 Networking Domain

The networking application domain covers all kinds of macroscopic communication devices. Standardization societies such as IEEE, ITU, and ETSI work out communication standards to achieve a high degree of interoperability. Additionally, the framework of the widely accepted ISO/OSI reference model [27] has been useful in providing a common terminology, stacking of communication services, and modularity of networking applications.

Concerning the variety of standards available for the respective ISO/OSI layers, this application domain follows an hour-glass scheme: A small set of networking layer standards in the middle of the ISO/OSI stack address a multitude of higher layer application standards as well as lower physical/link layer standards.

In principle, all different kinds of applications are characterized by their respective Quality of Service (QoS) requirements, which are condensed into set of service classes: Constant Bit Rate (CBR) traffic (e.g. telephony), Variable Bit Rate (VBR) real-time traffic (e.g. multimedia streaming), and Available Bit Rate (ABR) non-real-time traffic (file transfer).

Various efforts have been made to establish an integrated networking layer standard supporting all different service classes: the Integrated Services Digital Network (ISDN) was a first step into this direction. However ISDN is based on circuit switched communication and thus very inefficient for the increasing portion of bursty data traffic. The preceding Asynchronous Transfer Mode (ATM) employs packet switching to increase the resource utilization for non-CBR traffic. The dissemination of ATM has been hindered by the significant protocol overhead, which originates from the sophisticated signalling stack and flow-control mechanisms. This signallig is required to establish and maintain the state information related to the virtual channels and virtual paths. Today's de facto networking layer standard is given by the rather simplistic Internet Protocol (IP).

The variety of lower layer standards address specific physical networks: the core network communication backbone is predominantly established by Synchronous Optical Network (SONET) and Wave Division Multiplexing (WDM) based optical transmission. In the access network domain, a multitude of standards is available for Local Area Network (LAN) switching (Ethernet, FDDI, Token Ring), Wireless LAN (802.11a/b/g), and Wide Area Network (WAN) edge termination (analog/cable/xDSL/ISDN modems, telephony, access concentrators).

Looking at the SoC implementation complexity, the physical and link layer data rates of core network equipment are imposing demanding performance requirements. However the low flexibility of these standards allows for a hard-wired ASIC or even pure optical implementation. On the other side, higher application layers are only present in the terminal devices, so the relatively low to medium throughput requirements allow for a software implementation of the flexible and control dominated functionality.

In terms of SoC implementation complexity, the networking layer functionality constitutes by far the most challenging layer of the ISO/OSI reference model. Layer three multi-service access switches are considered as one of the potential killer applications for MP-SoC platforms, since they combine the physical wire speed throughput requirements with flexibility constraints imposed by the individual treatment of different service classes and application characteristics [28]. Advanced features like support for security sensitive applications in Firewalls or Virtual Private Networks (VPNs) further increase the processing requirements.

2.2 Multimedia Domain

The multimedia application domain subsumes the processing of all kinds of media data e.g. pictures, audio, video decoding, video pixel processing and 2D/3D graphics. Similar to the networking domain, a variety of standards enable the exchange of media data as well as device interoperablity. The advent of *digital* media processing has produced a multitude of standards, which realize different optima with respect to transmission bandwidth efficiency, processing requirements and quality. Table 2.1 summarizes computation, communication and memory requirements of typical multimedia standards [29].

Table 2.1. Characterization of Multimedia Applications.

application	*computation*	*communication*			*memory*
		in	*out*	*local*	
audio	100 MOPS	32-640 kbps	5 Mbps	5 Mbps	50 kb
MPEG2	4 GOPS	10 Mbps	120 MBps	240 MBps	8 MB
pixel	100 GOPS	360 Mbps	360 MBps	360 MBps	4 MB

Advances in processing capabilities and multimedia algorithms together with increased user expectations fuels a constant proliferation of new multimedia standards like digital audio decoding (AC3, OGG, MP3), video decoding (MPEG2, MEPEG4, H.263, H.264, DivX, quicktime), and 3D graphic processing (DirectX 9).

Apart from the multitude and dynamics of multimedia standards, a flexible implementation platform is also mandatory to meet demanding cost constraints of converging consumer electronics devices such as the Advanced Set-Top Box (ASTB). Here the processing and communication fabrics have to be shared among the multitude of supported multimedia applications to limit implementation cost.

2.3 Wireless Communications

The wireless communication application domain is characterized by an aggressive use of digital signal processing to maximize bandwidth efficiency. Again, a multitude of standards exists, each marking a local optimum in the multi dimensional parameter space spanned by implementation cost, mobility, power dissipation, and performance bandwidth efficiency. The statistic in figure 2.1 shows the numbers of changes to the UMTS standard over time to again emphasise the need for highly flexible embedded systems.

The multimedia and wireless communication domains are converging into a new generation of Personal Digital Assistant (PDA) or SmartPhone devices. So far PDAs run emaciated versions of typical desktop applications like organizer, info manager, text processors, spread sheets, presentations, or www browser. Recently, PDAs have started to support a huge variety of travel and fun related applications with much higher processing requirements, like e.g. localization, navigation, travel assistant, video camera, digital camera, picture editing, MP3

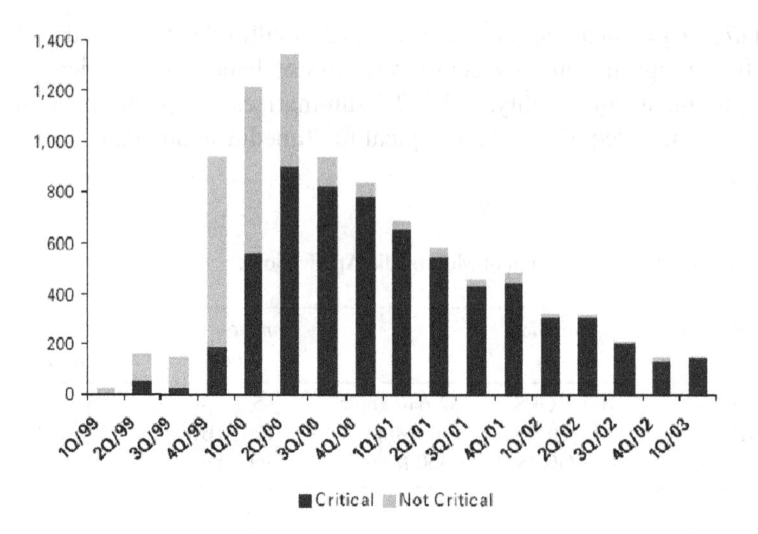

Figure 2.1. 3GPP Standard Changes

player, or games. Additionally, this kind of portable, multimedia enabled PDA devices are obliged to support multiple communication standards, both cable (USB, FireWire) and wireless (3G, WLAN).

2.4 Application Trends

The above considerations of the different embedded application domains with respect to SoC implementation can be summarized into the following set of common trends:

- New features and value added services, together with the heuristic logarithmic law of usefulness [30], lead to *exponentially increasing processing performance and communication requirements.*
- The standards become more dynamic and sophisticated and are introduced more rapidly. This calls for *high flexibility* of the SoC implementation to meet the resulting time-in-market as well as time-in-market requirements.
- For mobile applications as well as for cost sensitive consumer electronic devices, *energy efficiency* becomes the prevailing cost factor.

Heterogeneous Multi-Processor SoC (MP-SoC) platforms are generally believed to meet the above mentioned conflicting performance, flexibility and energy efficiency requirements of demanding embedded applications. The heterogeneity of future SoC implementations is driven by the heterogeneity of the embedded applications, where each part of the application has an inherent optimal implementation. Hence, in the course of an MP-SoC platform design the *partitioning* of a specific application is a task of major importance.

2.5 First Order Application Partitioning

A first order partitioning into a control dominated domain and a data dominated domain can be applied to every embedded application, no matter which application domain is considered. This first order partitioning has major influence on both the target processing and communication elements as well as on the appropriate design methodology. Figure 2.2 shows control- and data-plane processing tasks for selected example applications.

Application	Data-Plane Processing	Control-Plane Processing
IP forwarding with QoS	queuing, scheduling, routing, classification, en-/decryption	policy applications, network management, signaling, topology management
Advanced Set-Top Box (ASTB)	audio decoding, video decoding, 3D graphic processing	configuration management, user interaction
wireless PDA	UMTS/WLAN modem	Personal Information Management (PIM), office applications, games,

Figure 2.2. Control-/Data-Plane Processing for Selected Example Applications

Control-Plane Processing

Control-plane processing is characterized by moderate performance requirements, but on the other hand comprises huge amounts of functionality calling for maximum flexibility. Example control-plane processing tasks in the networking application domain are, e.g. policy applications, network management, signaling, or topology management.

The control plane functionality is usually developed using an architecture agnostic, software centric Integrated Design Environment (IDE) and state-of-the-art software engineering techniques like Object Oriented Programming (OOP) using the Unified Modeling Language (UML) [31], C++ [32], or Java [33]. To increase the reuse of the control plane Software across multiple MP-SoC platform generations, the Hardware dependant Software (HdS) portions are wrapped into a stack of middleware, Real Time Operating System (RTOS), and device driver layers [34, 35].

The huge amount of functionality and little inherent parallelism of control plane processing tasks usually prohibits the explicit specification of Task Level Parallelism (TLP). Thus, in order to gain performance the designer relys on fine grain Instruction Level Parallelism (ILP) to be extracted by a VLIW compiler or by a superscalar processor architecture.

Data-Plane Processing

Data-plane processing is characterized by computationally intensive data manipulations performed at high data rates, thus demanding high processing and communication performance. Additionally, rapidly evolving standards in all application domains impose increasing flexibility constraints. Example data-plane processing tasks in the networking application domain are e.g. queuing, scheduling, routing, classification, or en-/decription.

The performance requirements of networking, multimedia and wireless communications applications can only be reached by aggressively exploiting the abundant inherent parallelism available in the data-plane processing tasks:

- The functionality can be straightforwardly partitioned into a set of loosely coupled tasks with well predictable or even cyclo-stationary execution timing.
- A well confined data set is associated with a single activation of an individual task. Additionally, the data sets associated with successive activations of an individual tasks are mostly independent.

These spatial and temporal properties with respect to second order task partitioning and data dependency can already be identified during the algorithm development stage and lead to an identification of coarse grain TLP. This application inherent TLP enables the concurrent and parallel execution on MP-SoC platforms.

Chapter 3

CLASSIFICATION OF PLATFORM ELEMENTS

Current SoC architectures still very much follow the *System-on-a-Board-that-happens-to-be-on-a-Chip* paradigm [36]. That is to say, the processing of embedded applications is implemented as a mix of dedicated hardwired logic blocks and general purpose processors executing the embedded Software. The on-chip communication is mostly based on shared bus architectures, which are quite similar to the tristate buses known from the Printed Circuit Board (PCB) world.

This kind of PCB inspired SoC architectures fail to deliver the performance, energy efficiency and flexibility required by the demanding embedded applications discussed in the previous chapter. Instead, future SoC architectures will be assembled from a huge variety of processing kernels and interconnect networks, which are individually configured and specialized for the target application.

The first part of this chapter briefly introduces the most important architectural metrics. Based on these metrics the main body of this chapter classifies processing elements as well as on-chip communication architectures. Finally, the discussion of embedded applications and SoC architectures is summarized to derive the requirements for the next generation SoC design methodology.

3.1 Architecture Metrics

This section introduces a set of macroscopic metrics for the classification and evaluation of architectural elements.

Cost. The Cost of an embedded architecture is separated into the Non Recurrent Engineering (NRE) cost for the initial design and recurring chip fabrication cost. The major NRE cost factor is caused by the design effort for HW and

SW development, but also comprises the fabrication of the initial mask set. Typical NRE cost of an 90 nm technology SoC is the order of 10-100 Million USD design effort and 1 Million USD per mask set. The fabrication cost for a given technology node is determined by the silicon die area and the packaging, which in turn is determined by the number of pins and the power dissipation requirements.

Performance. The Performance of both computational and communication architectures is further classified into *latency* and *throughput.* Latency denotes the absolute *time* passing between the start and completion of a task, whereas throughput in general refers to the number of accomplished *tasks per time.* Communication throughput is therefore measured in transferred bits per second (bps). On the other hand, throughput of programmable processing elements is measured in Millions Instructions Per Second (MIPS). Despite the wide usage of the MIPS metric, it is not always meaningful to characterize the expected application performance for non-RISC processor architectures.

Power Dissipation. measured in Watt denotes the *energy per time* required to operate an embedded system and is an architecture metric of growing importance. First, the battery lifetime of mobile devices immediately depends on the energy consumption. Second, the packaging cost depends on the heat dissipation properties, which in turn depends on the power consumption. As shown below, striving for low power and energy consumption constitutes the key driver for architecture differentiation of embedded SoC platforms.

Computational Efficiency. is derived from performance and power consumption. It characterizes the efficiency of a given architectural element with a single value. Computational efficiency of programmable architectures is predominantly measured in MIPS/Watt. Since the inaccuracy of the MIPS metric propagates into the MIPS/Watt metric, computational efficiency — especially in the context of battery enabled applications — is alternatively measured in energy consumption per task.

Flexibility. is related to the effort to change the functionality of a given architectural element. In contrast to the previous metrics, flexibility can be hardly measured in an accurate way. Nonetheless, in the context of rapidly evolving functionality and standards of embedded applications, architectural flexibility is of major importance to achieve both decreasing time-to-market as well as increasing time-in-market.

3.2 Processing Elements

In general, a processing element (PE) provides the computational resource to execute a given portion of the application. The type of a PE has traditionally been selected along a black-and-white performance/flexibility trade-off: Application Specific Integrated Circuit (ASIC) architectures are hardwired implementations of a fixed application set providing highest possible performance close to the inherent silicon capabilities. On the other hand, programmable PEs are controlled by an instruction stream in a highly flexible way.

The rather poor performance of programmable PEs has ever fueled computer architecture research towards parallelizing the execution of instructions. Early efforts in parallel computer architectures are classified by Flynn [37] according to the deployment of control- and data-level parallelism:

- **SISD**, Single Instruction Single Data refers to the traditional von-Neumann kind of computer architectures, which sequentially execute a single instruction stream on a single processing resource.
- **SIMD**, Single Instruction Multiple Data vector processing machines perform a single instruction on multiple data items in parallel. SIMD processing is still heavily used in state-of-the-art architectures for embedded DSP and graphic applications to exploit inherent data-level parallelism (DLP).
- **MIMD**, Multiple Instruction Multiple Data denotes the traditional homogeneous multi-processor type of architectures employed in scientific supercomputers like Cray T3E, or the NEC Earth Simulator.
- **MISD**, Multiple Instruction Single Data is a rarely encountered class of architectures, which exploit temporal ILP by setting pipeline stages and executing several instructions simultaneously, e.g. vector pipelining in CRAY-1.

Complementary to this traditional classification, more recent performance enhancement strategies are discussed in the following sections.

3.2.1 Processing Element Trends

Enabled by the constant progress in silicon technology, new computer architecture features have been invented to incrementally improve the application throughput of programmable architectures:

- **Superpipelining** uses deep execution pipelines to increase the clock frequency.
- **Superscalarity** employs parallel functional units and complex dispatcher architectures to dynamically extract Instruction Level Parallelism (ILP).

- **Very Large Instruction Word (VLIW)** architectures execute several statically scheduled instructions on parallel functional units, hence the effort for ILP extraction is moved into the compiler.

- **Hardware Multi-Threading (HW-MT)** architectures [38, 13] are able to concurrently pursue two or more threads of control by providing separate register resources for each thread context.

- **Domain Specific (DS) Instruction Set** tailors the programmable PE to a specific application domain by providing specialized functional units. DS processor examples are Digital Signal Processors (DSPs) employed in multimedia and wireless communications, or Network Processing Units (NPUs) for networking applications.

The applicability of the above listed performance improvement techniques depends on the considered set of target applications. Superpipelining and Superscalarity are heavily used in high performance General Purpose Processor (GPP) architectures to increase single thread performance of arbitrary applications on the vast expense of silicon area and power dissipation.

On the one hand, embedded applications are severely energy and cost constrained, but still have significant performance and flexibility requirements. The most promising approach to jointly optimize flexibility and performance is to exploit coarse-grain TLP instead of ILP [39] and map the loosely coupled tasks to individually optimized PEs. This kind of embedded PEs mostly rely on the more power aware performance optimization techniques, like VLIW, multi-threading and a domain specific or even application specific instruction set [11].

3.2.2 Parallel Multi Processing

The MIMD kind of control parallelism plays an increasing important role in embedded SoC architectures, because parallel execution of specialized PEs offers a chance for improving application performance without sacrificing power efficiency.

Homogeneous Multi-Processing. refers to the multiple instantiation of identical PEs and thus corresponds to a single chip implementation of the MIMD principle. On the one hand side, homogeneous multi-processing of general purpose embedded micro controllers is considered to achieve the performance scaling required for control-plane processing portion of embedded applications [40].

On the other hand, homogeneous multi-processing is also found for data-plane processing in domain specific MP-SoC platforms, where the identical instruction set of the PEs is tailored to a certain application domain.

Heterogeneous Multi-Processing. employs multiple PEs, which are individually tailored to a certain task or task set. This kind of dedicated optimization is only applicable for the data-plane processing portion of the application, which allows for a manual and static task allocation. The high degree of specialization in heterogeneous multi-processing further optimizes computational efficiency for a well defined set of target applications at the expense of generality.

3.2.3 Concurrent Multi Processing

Parallel execution described in the previous section requires multiple computational resources, hence more than one task can be active at the same point in time. On the other hand, *concurrent* execution denotes the interleaved processing of several tasks on a single resource, such that at any time only one task can be active.

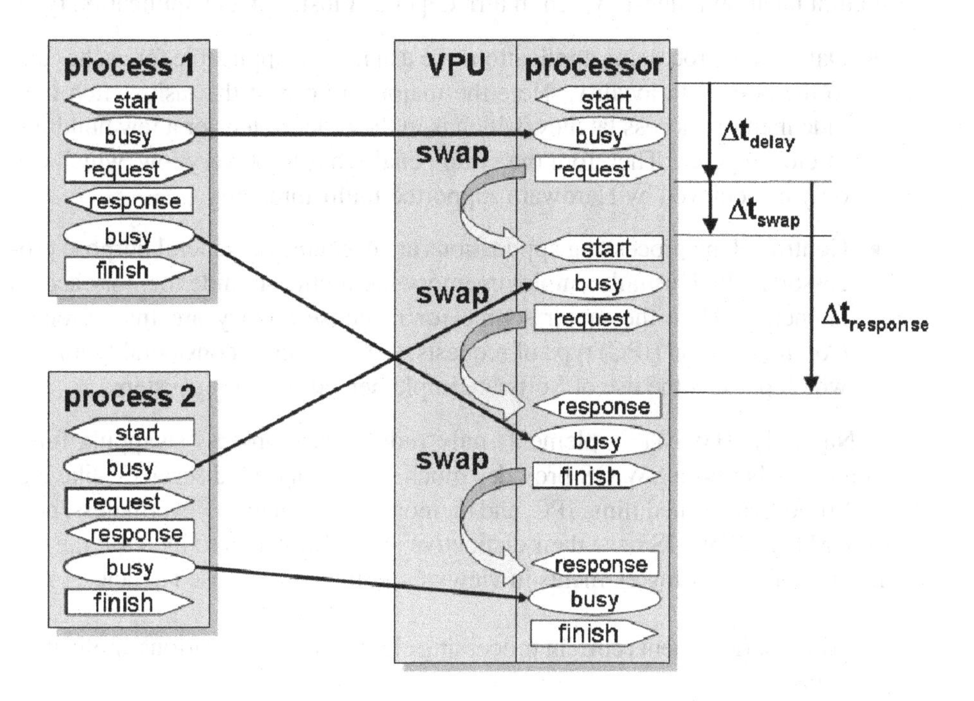

Figure 3.1. Multi-Threaded Processing Element

The benefit of concurrent execution is depicted in figure 3.1, where two tasks are mapped to a single processing element. Both tasks are divided into two processing portions, which are separated by a communication request. After Δt_{delay} the processing of the first portion is finished and the task is blocked for $\Delta t_{response}$ until the request is accomplished. Instead of wasting the processor resource during this period, the processor context is swapped to the second task by a scheduler. Hence the utilization of the processor is increased and the request latency is hidden.

The task scheduler can be implemented either in Hardware or in Software. Hardware Multi-Threaded (HW-MT) processor architectures provide dedicated hardware support in terms of multiple register files to enable context swapping within a single cycle. In contrast Software Operating Systems (SW-OS) explicitly save the register context of the preempted process and then load the restored process. In this case, the associated context swap penalty Δt_{swap} is in the order of tens or a few hundreds of clock cycles.

The choice between HW-MT and SW-OS is determined by the time scale of Δt_{delay}, $\Delta t_{response}$ and the process swap penalty Δt_{swap}. Clearly, Δt_{swap} has to be much smaller than $\Delta t_{response}$, otherwise the processor utilization gain would disappear. This time scale is mostly determined by the type of communication request, which in turn depends mostly on the application type:

- Data-plane processing applications are usually not applicable for caches due to the poor data locality. Here the major purpose of the task switch is to hide memory access latency, which is in the order of tens or a few hundreds of clock cycles. Therefore the swap penalty has to be very low, which can only be achieved by Hardware supported multi-threading.
- Control-plane processing applications are executed on general purpose processors, which usually employ memory hierarchies to hide memory access latencies. Here the major source for response latency are Inter Process Communication (IPC) type of requests. IPC latency is considerably longer, which permits the use of Software implemented operating systems.

Naturally, HW-MT implements only rudimentary process swapping functionality whereas a SW-OS provides much more elaborated services, like e.g. real-time aware scheduling, IPC, and memory management. Nevertheless, both HW-MT and SW-OS have their distinctive application area to increase the utilization of processing elements in view of significant response latencies.

Table 3.1 documents current processing element trends in various application domains.

3.3 On-Chip Communication

This section classifies known and emerging communication architectures. For this discussion the same basic cost, performance, power, and flexibility metrics already introduced in section 3.1 apply. Additionally, Quality of Service (QoS) metrics known from the networking application domain[1] are of

[1]please refer to section 2.1 on page 9

Table 3.1. Example Processor Architectures.

Processor	*Characteristics*	*Reference*
Micro Controller		
ARM	popular RISC processor, trend towards moderate superscalarity and deeper pipelining	[41]
MIPS	popular RISC processor, trend towards co-processor, user defined instructions, Application Specific Extensions (ASE)	[42]
Digital Signal Processor SoC		
Sandbridge SB3010	4 way homogeneous multi-processor, SIMD vector DSP unit, 8 way HW-MT, RISC based integer unit	[16, 43]
Network Processing Units		
Intel IXP 2400	8 way homogeneous multi-processor, RISC-like micro engines with 8 way HW-MT, XScale micro-controller	[44]
Agere PayloadPlus	heterogeneous multi-processor, 64 way HW-MT Fast Packet Processor (FPP), VLIW Routing Switch Processor (RSP)	[45]
AMCC nP3700	3 way homogeneous multi-processor, RISC-like nPcores with 24 way HW-MT	[46]

increasing importance to manage complex on-chip traffic. In the face of the rapidly growing number of processing elements, also the *scalability* of the communication architecture gains growing attention.

3.3.1 Bus Architecture

The bus based on-chip communication paradigm is derived from the Printed Circuit Board (PCB) domain such as the VME (Versa Module Eurocard bus [47]) and PCI (Peripheral Component Interconnect [48]). Due to the easy programming model, high flexibility and abundant availability of Intellectual Property (IP), this concept is clearly advantageous for today's small and medium scale embedded systems, where a small number of blocks exchange moderate amounts of data.

Typical state-of-the-art bus systems as depicted in figure 3.2 implement a master-slave communication scheme, where active initiators along with passive target modules are hooked to a shared communication medium [49]. Typical masters are processors, DMA controllers or autonomous ASIC blocks, whereas typical slaves are memories, co-processors and other peripherals.

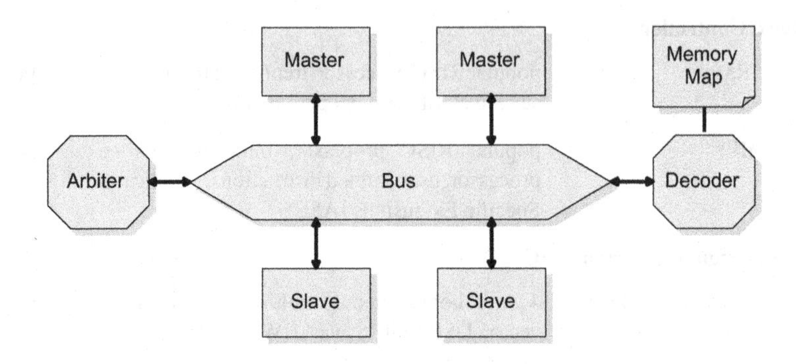

Figure 3.2. Schematic Bus System

Further components of a bus system are arbitration and decoder units. The bus arbiter grants the access to the communication medium to one of the competing master modules. The decoder activates the target module based on the actual address and the address map, which maps the target modules into the bus address space.

The following sections enumerate typical bus features and discuss the merits and shortcomings of bus based communication.

On-Chip Bus Characteristics

Modern bus systems provide a huge variety of design parameters, which can be tailored to the considered application in order to reduce bus contention and to meet the respective performance requirements [50]. **Bandwidth** is the premier performance metric and denotes the maximum transfer capacity of the bus. The available bandwidth is measured in bits per second and corresponds to the number of parallel data wires divided by the bus clock period.

Pipelining is a well known technique to improve the communication throughput. Like in processing elements, the clock frequency is limited by the critical path. Hence, inserting an additional pipeline stage into the critical path allows a higher clock frequency and thus yields a higher communication bandwidth. Since the address decoder is usually integral part of the critical path, bus transactions in high performance buses are executed in separate address and data stages.

Burst modes further improve communication throughput for the linear access of subsequent addresses by a single master. In this case the address counter is incremented automatically and the next data item is transferred with every cycle without renewed arbitration.

Unidirectional data links distinguish on-chip buses from most on-board buses [51]. The latter are based on tristate data wires to maximize the utilization of expensive on-board wires.

Hierarchy refers to the fact, that common bus systems separate high performance from low performance communication by providing two buses with different speed characteristics.

Multilayer bus architectures provide dedicated point-to-point connections between distinctive initiators and targets to eliminate bandwidth bottlenecks. The required de-multiplexer at the initiator side is called input stages, the respective target multiplexer is called output stage.

Crossbar bus architectures provide multiple parallel resources between initiators and targets to significantly improve the traffic throughput. The degree of parallelism may vary from partial crossbar to full crossbar architectures, where the latter provides an individual resource for each connected target.

Arbitration can be based on various algorithms, ranging from simple round robin, via fixed, configurable or dynamic priority schemes to static or dynamic Time Division Multiple Access (TDMA) schedulers. Even more advanced algorithms are known to further improve the quality of service.

Locking of a bus by a single master is a necessary feature to support read-modify-write kind of semaphore operations. This feature is required by most micro-controller architectures, which run operating systems.

Split transaction buses allow the master to issue multiple requests without waiting for a response, i.e. request and response are separated [52].

Out-of-order execution further improves the bus throughput by reordering the sequence of responses, depending on the availability of the slave component. This feature requires advanced state-machines in the master modules to cope with non-deterministic sequence of responses.

As demonstrated by the example bus architectures in table 3.2, available and emerging bus systems more or less offer this comprehensive set of architectural choices. This already highlights the current challenge for the system architect to conceptualize an optimal communication architecture for a given application.

Table 3.2. On-Chip Bus Architectures.

Bus Architecture	*Characteristics*	*Reference*
commercial		
ARM AMBA	popular hierarchical bus system: multi-master Advanced High-performance Bus (AHB) with central priority based arbiter, pipelining, burst transfers, split transactions, multi-layer single master Advanced Peripheral Bus (APB)	[53]
IBM CoreConnect	hierarchical bus system similar to AMBA: high performance Processor Local Bus (PLB), multi-layer capabilities by means of PLB Crossbar Switch (PCS) low performance On-Chip Peripheral Bus (OPB)	[54, 55]
STBus	highly configurable bus system, 3 hierarchy levels (peripheral, basic, advanced), configurable arbitration scheme (priority based, latency based, LRU), out-of-order execution, separate request-response resources, crossbar bus, internal buffers	[56, 57]
Sonics μNetwork	Open Core Protocol (OCP) interface, guaranteed bandwidth through Time-Division Multiple Access (TDMA) and Round Robin (RR) arbitration mechanisms	[22, 58]
academic		
Lotterybus	probabilistic arbitration scheme	[59]
HIBI	priority and time-slot based distributed arbitration	[60]

Drawbacks

Despite their current popularity, the shared bus communication paradigm increasingly fails to cope with communication infrastructure requirements of large scale MP-SoC platforms:

Physical Issues. Current bus architectures are implemented using a standard cell based semi-custom implementation flow. Hence, the transmission wires are not physically optimized, which in the current and coming semiconductor technology nodes leads to timing closure issues and unreliable communication links. Examples of physical effects are crosstalk noise, electromagnetic interference, and radiation-induced charge injection [61, 62, 7].

Synchronous Design. Most current bus architectures require, that all connected modules are situated in a single clock domain. Due to the parasitic capacities of long bus wires, strong driver transistors are necessary to achieve timing closure. This in turn leads to the fact, that already today the on-chip communication infrastructure is the major origin of power dissipation. Future SoC designs will follow the Globally Asynchronous Locally Synchronous (GALS) [63] paradigm, thus chip-wide wires will span multiple clock domains, which disqualifies bus architectures as the future chip-level transport mechanism.

Traffic Management. Due to the rather simple arbitration mechanisms, shared buses provide only rudimentary traffic management support. Since the communication pattern highly depends on the spatial and temporal execution of the application tasks, meeting the individual QoS requirements like throughput, jitter, or ordering of the respective tasks is very challenging. This also causes the poor scalability of bus-based communication infrastructures, since every change in the traffic profile of one part of the application and every additional component influences the other parts and requires renewed balancing of the bus architectures.

Interoperability. Although simple standard peripherals, like DMA, IRC, or memories are available for respective bus systems, it is a tedious and error-prone task to adapt complex IP blocks to a specific bus architecture. So far efforts to create standard bus interfaces, like e.g. VSIA [52] or OCP-IP [22] have not been successful.

3.3.2 Network-on-Chip Architectures

Researchers in academia and industry have conceived alternative on-chip communication concepts to cope with the limitations of shared bus architectures. These efforts have recently been subsumed under the *Networks on Chip* (NoC) design paradigm [7, 64]. The NoC paradigm aims to replace current adhoc wiring of IP blocks with a disciplined approach, where full-scale on-chip networks provide communication services according to the ISO/OSI reference model [65, 66]. By that the manifold problems in on-chip communication like signal integrity issues, link reliability, or Quality of Service (QoS) are separately resolved on the respective OSI layer.

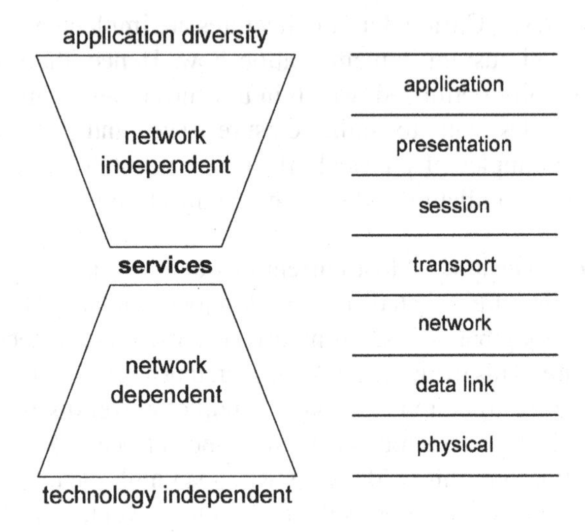

Figure 3.3. ISO/OSI Reference Model Services [67]

In the context of on-chip network, the four lower layers of the ISO/OSI reference model are of interest:

- The **Physical Layer** deals with the electrical aspects of the data transmission, like e.g. signal voltages, clock recovery, and pulse shape. In the future, the physical layer of on-chip networks may incorporate transmission technology known from cable modems or even the wireless communication domain [68] like synchronization, channel estimation, channel coding/decoding to cope with unreliable transfer channels.

- The **Data Link layer** provides a reliable data transfer over the physical link. This may include error detection by means of block codes and error correction mechanisms like Automatic Repeat Request (ARQ) or Forward Error Correction (FEC).

- The **Network Layer** implements the arbitration algorithms, buffering strategies and flow-control mechanisms. By that the networking layer has dominant impact on the performance and functional behavior of network. These aspects are further elaborated in the remainder of this section.

- **Transport Layer** protocols establish and maintain end-to-end connections. Among other things, the transport layer manages rate-based flow control, performs packet segmentation and reassembly, and ensures message ordering. This abstraction hides the topology of the network, and the implementation of the links that make up the network.

Many results from research on macroscopic computer networks can be employed to solve the on-chip communication issues. As depicted in table 3.3, Network-on-Chip architectures impose a specific set of implementation constraints, which differ significantly from macroscopic computer networks [7].

Table 3.3. Network Comparison

Aspect	*Macroscopic Networks*	*Networks on Chip*
Generality	general purpose, modularity, standards (IP, ATM, Ethernet)	design time specialization, application specific
Wires	long transmission wires, millions of bits on-the-fly	short wires, < 10 bits on-the-fly
Flow-Control	sophisticated flow control	simple back-pressure
Memories	large, cheap off-chip DRAMs for packet payload	on-chip memory dominating cost factor

The challenge in the development of Network-on-Chip architectures is to combine the know-how from *both* the networking and VLSI domain. But also the users of on-chip networks have to understand basic networking principles: First the system architect has to specify design time parameters of the selected NoC architecture like topology, buffer sizes, arbitration algorithm. Later the platform programmer has to configure runtime parameters like priorities, routing tables, buffer management thresholds to take advantage of the capabilities.

The following paragraphs introduce transport and network layer principles, which are important in the context of on-chip networks.

Transport Layer Services

As depicted in figure 3.3, the transport layer is the first to provide services, which are independent of the implementation of the network [69]. This enables the platform programmer to develop embedded software independently from the interconnect architecture. This is a key ingredient in tackling the challenge of decoupling the computation from communication [66],[70]. By that, interaction with the network becomes deterministic, rather than prognostic or reactive like in today's bus based communication architectures.

For complex multi-hop networks it is difficult to provide uniform Quality of Service (QOS) guarantees like lower bandwidth bounds, or packet ordering for the complete on-chip traffic. To combine high resource utilization with high QoS requirements of certain traffic types, researchers in the field of computer networks distinguish guaranteed services and best effort service classes [71].

Two basic service classes [8] are known from research on macroscopic computer networks:

- **Guaranteed Services** require resource reservation for worst-case scenarios. This can be rather expensive since guaranteeing the throughput for a stream of data implies reserving bandwidth for the peak throughput, even when its average is much lower. As a consequence resources are often underutilized.

- **Best-effort Services** do not reserve any resources, and hence provide no guarantees. Best-effort services utilize resources well due to the fact that they are typically designed for average-case scenarios instead of worst-case scenarios. They are also easy to configure, as they require no resource reservation. The main disadvantage here is the unpredictability of the effective performance.

Network Layer Mechanisms

The ISO/OSI networking layer is implemented by the routing nodes of the NoC. Router based network implementations can be classified according to the following categories:

- **switching mode:** Two switching modes can be distinguished: circuit switching and packet switching.

  - **circuit switching:** In a circuit-switched network connections are set up by establishing a conceptual physical path from a source to a destination. Links can be shared between two connections only at different points in time, by using the time-division multiplexing (TDM) scheme.
  - **packet switching:** In a packet-switched network the data is divided into packets and every packet is composed of a header and the payload. The header contains information that is used by the router to switch the packet to the appropriate output port.

- **routing mode:** applies only to **packet-switched** networks and defines the way packets are transmitted and buffered between the network nodes. These are [72]:

- **store-and-forward:** Sn incoming packet is received and stored entirely before it is forwarded to the next node.
- **wormhole routing:** An incoming packet is forwarded as soon as the packet header is evaluated and the next router guarantees that the complete packet will be accepted. In case the next hob is blocked, the packet tail potentially blocks other resources.
- **virtual cut-through:** An incoming packet is forwarded as soon as the next router guarantees, that the complete packet will be accepted. In case the next hob is blocked, the packet tail is stored in a local buffer.

- **Queuing:** Buffering strategies can be distinguished by the location of the buffers inside the router. Input queuing and output queuing and variants of them can be distinguished [73](see figure 3.4). In the following, N denotes the number of bi-directional router ports.

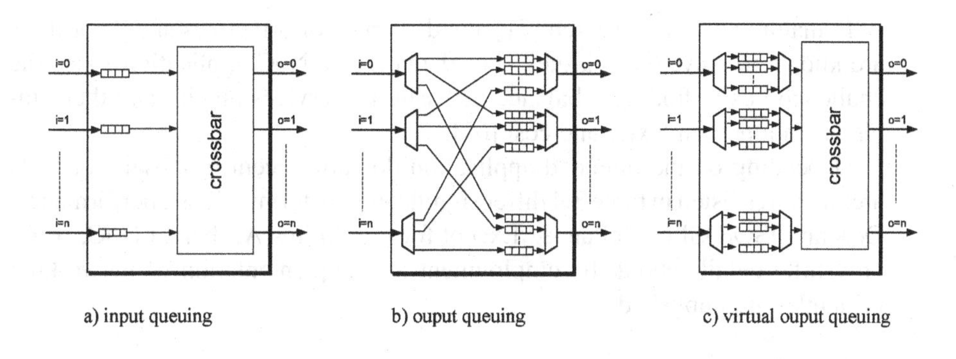

Figure 3.4. Queuing Schemes

- **input queuing:** In input queuing a router has a single input queue for every incoming link. Input Queuing suffers from the so-called head-of-line blocking problem, i.e. the router utilization saturates at about 59% [74], resulting in weak link utilization.
- **output queuing:** In output queuing there are N output queues for every outgoing link resulting in N^2 queues. Although this approach yields optimal performance, the costly N^2-fold storage and wiring effort prohibits the implementation of output queuing for a large number of ports
.
- **virtual output queuing:** Virtual output queuing (VOQ) [74] combines the advantages of input queuing and output queuing and avoids the head-of-line blocking problem. In this technique each input port maintains a separate queue for each output port. One key factor in achieving high performance using VOQ switches is the scheduling algorithm.

- **congestion control:** Packet switched networks without mechanisms for bandwidth reservation may run into resource contention and subsequent buffer overflow. Several solutions prevent packets from entering until contention is reduced:
  - **packet discarding:** simply drops packets in case of buffer overflow.
  - **credit based flow control:** packet loss is prevented in a deterministic way by either signaling congestion via separate wires (back-pressure) or the receiver regularly informs the sender about the available buffer space (window).
  - **rate based flow control:** the sender gradually adjusts the traffic generation rate in response to control flow messages from the receiver. Rate based flow control has to be implemented by the transfer layer and potentially suffers from instability due to long control loops.

In mature network research [71] a wide variety of principles and algorithms are known to solve these issues. For the concrete NoC application space the challenge lies in finding a balance between the services quality and their implementation complexity and cost [67].

Depending on the intended application domain, recent proposals for NoC architectures listed in table 3.4 differ significantly in terms of cost, performance, QoS and ISO/OSI compliance. Except for the Philips AEthereal NoC, so far only little details about the employment and implementation of networking principles are published.

3.4 Summary

This section summarizes architectural trends from the previous sections to set the stage for the discussion of appropriate system level design methodologies in the following chapters.

Concerning the processing elements, the conflicting requirements for performance, power efficiency and flexibility drive the diversification into highly application specific solutions. Especially increasingly complex and dynamic application standards foster a clear trend towards programmable processing elements. Computer architecture features like SIMD, VLIW, superpipelining, and hardware multi-threading are applied to exploit application inherent instruction-, data-, and task-level parallelism.

On the other hand, the real challenge of system architecture design will shift from computation to communication. Here, the high speed processor bus architectures for local communication will be complemented by full-scale on-chip networks to handle the global traffic. Numerous competing communication architectures, topologies and protocols are at disposal and more are proposed or under development. Even after the decision on the type of communication IP,

Table 3.4. Network-on-Chip Architectures

NoC architecture	*Characteristics*	*Reference*
Arteris	homogeneous network based on scalable, parameterizable packet switches, multi-purpose data packets, traffic management capabilities, GALS approach, bridges to standard buses	[9]
Philips AEthereal	packet switched network, wormhole-routing, integrated combination of guaranteed bandwidth and best effort services	[8, 67]
NOC	packet switched network, two-dimensional Manhattan grid of 5x5 switches	[75]
PROPHID	time-space-time division 3 stage circuit switched network, static bandwidth allocation	[76]
SPIN	packet switched network, fat tree topology, wormhole routing, input queuing, credit-based flow control	[77]
STNoC	packet switched router with support and best-effort and QoS traffic, 'Spidergon' topology	[78, 79]

the designer still has to investigate the design space spanned by a single communication technology, like e.g. topology varieties, protocol options, priority schemes, or hardware parameters like bitwidth, clock cycle, and queue length.

To overcome design complexity, the *virtualization* of architectural resources enables 'divide-and-conquer' like temporal and spatial mapping of the application tasks:

- embedded control-plane processing tasks are executed in the user space the Real Time Operating System (RTOS), so the individual processing resource requirements are satisfied by the real-time task scheduler.
- embedded data-plane processing tasks are executed on HW multi-threaded processing elements to automatically hide communication latencies and thus increase the utilization of the computational resources.
- global communication of control- and data-plane processing elements is performed by elaborated on-chip networks, so the quality-of-service requirements of all tasks can be individually configured.

Thus, virtualization of the MP-SoC communication and processing resources is considered the most promising approach to enable the mapping of arbitrarily complex applications to heterogeneous MP-SoC platforms [17]: However, this *divide-and-conquer* approach enables the individual configuration of communication and processing resource requirements for each of the application tasks. On the other hand, virtualization of the shared resources foster global optimization of performance and utilization.

Still the major burden is laid on system architects to first employ and take advantage of high-volume asymmetric embedded processing elements as well as conceptualize an advanced heterogeneous on-chip communication infrastructure. Second, the temporal and spatial mapping of the application to the heterogeneous MP-SoC platforms of course has to meet the performance requirements, but should also render a high and balanced utilization of the shared communication and processing resources.

The subsequent chapter will discuss state-of-the-art methodologies and tools addressing the emerging complexity crisis of MP-SoC design and application mapping.

Chapter 4

SYSTEM LEVEL DESIGN PRINCIPLES

The traditional design flow of integrated circuits is separated into an application phase and a largely decoupled implementation phase. The transition from application to implementation is performed by means of a specification document written by the application team and system architecture specialist.

This ad-hoc and informal approach has always imposed a number of problems during the subsequent implementation phase.

- The ambiguity of the informal specification document leads to misinterpretations and implementation errors.
- The lack of reliable performance information before the implementation often causes an over- or under-provisioning of processing and communication resources.
- The quality of results mainly depends on the intuition and experience of the system architect.
- The manual creation of the verification environment requires significant effort and again represents a potential source of inconstistencies with the original design intend.

These disadvantages become overwhelming with the advent of the MP-SoC area, where the system architecture is far beyond what can be specified on paper. Attempts to bridge this gap applying tool automation have failed, either due to the poor quality of results or due to the restrictive input specification formalism.

Today an intermediate design phase called Electronic System Level (ESL) is emerging as the solution to the system complexity problem. In this phase the application is jointly considered with the system architecture to find a feasible and cost effective application to architecture mapping.

The declared goal of ESL design is to increase the engineering productivity and quality of results during the specification of the MP-SoC platform architecture and application mapping. *Orthogonalization of concerns* is generally believed to be the key ingredient for conquering the complexity by separating parts of the design process [34].

However, the exact contour of ESL Design is by far not mature and a lot of different approaches exist in industry and academia. This chapter introduces general concepts of platform based design, design phases, abstraction levels, and models of computation. Together with the properties of future MP-SoC platforms derived in chapter 3, these concepts provide the foundation for the succeeding discussion of existing approaches to SLD.

4.1 The Platform Based Design Paradigm

Compared to the traditional ASIC design era, Platform Based Design (PBD) is generally understood as a new design paradigm to cope with the complexity and the economics of the emerging billion-transistor System-on-Chip era. Especially the huge NRE cost prohibits the development of low volume ICs for a single application. In this context, the term Platform Based Design is currently holding two different meanings.

The first definition is rather architecture centric and declares the goal of PBD to greatly increase the flexibility by the employment of programmable or reconfigurable processing elements:

> We define platform-based design as the creation of a stable microprocessor-based architecture that can be rapidly extended, customized for a range of applications, and delivered to customers for quick deployment. Source: Jean-Marc Chateau, ST Microelectronics.

Another, more generalized understanding puts the design process itself into focus and defines PBD as a *meeting in the middle* approach, where the space of addressed applications meets the space of possible architectures in a unified platform API[1].

> The general definition of a platform is an *abstraction layer in the design flow* that facilitates a number of possible refinements into a subsequent abstraction layer in the design flow. [80, 34, 81, 82]

By the first definition, a platform can be considered as a synonym term for an MP-SoC implementation, whereas according to the latter definition, an MP-SoC implementation is merely a platform instance.

Apart from this subtleness in terminology, the trend towards flexible MP-SoC platforms has already been motivated in the previous chapter. Concerning the platform design process discussed in this chapter it is generally agreed, that

[1] API: Application Programmers Interface

in one way or the other heterogeneous MP-SoC platforms will be defined in a successive refinement of abstraction levels.

4.2 Design Phases

According to the *multi-level SoC design* approach proposed by Magarshak and Paulin [83], the overall MP-SoC design process can be separated into multiple, almost orthogonal phases:

- The **functional phase** is performed by application specialists and completely agnostic to architectural considerations. This phase includes the embedded SW development of the control-plane portion of the application as well as data-plane algorithm development. As discussed below, the latter is carried out using highly application domain specific tools and methodologies.

- The **MP-SoC platform phase** covers all designs tasks, which which have to be performed under consideration of the full functional and architectural complexity the MP-SoC platforms. This comprises for example the specification of the system-architecture, the mapping of the application onto the MP-SoC platform, but also the development of the hardware dependant Software layers.

- The **high-level IP creation phase** deals with the design of processing elements (RISC, DSP, MCU, ASIPs), on-chip interconnect technologies (busses, NoC), domain specific standard I/O (PCI-variants, SPIx variants, HyperTransport, I2C, FireWire, QDR, etc.), as well as the creation of well defined ASIC IP blocks (e.g. an MPEG4 video codec). This phase is not completely orthogonal to the functional phase, since the design of application specific processing elements and communication IP indeed depends on the considered application.

- **Semiconductor technology and basic IP creation phase** covers standard cells, I/O, memories and the basic technology processes supporting them. The trend here is for more heterogeneous technologies, combining embedded DRAM, embedded Flash, mixed-signal BiCMOS, RF, and analog. This phase is of less interest in the context of this book.

These tasks are not performed in a sequential top-down fashion. Especially IP (both high and low level) is usually created independent from a particular platform instance. Exceptions from this is the specialization of IP for a particular application, like e.g. the development of an Application Specific Instruction-set Processor (ASIP), or an Application Specific Bus Architecture.

Also, there is no one-to-one correspondence between design phase and abstraction level. Instead, the design phases cover several abstraction levels.

The topic of book is clearly related to the MP-SoC platform phase. Therefore the different design tasks of this phase shall be investigated in more detail.

First of all there is a need to represent the results of the functional phase as a well defined *application model*. In the course of this task the selected algorithms are decomposed or assembled into a set of tasks and communication channels. Also the missing functionality for the control and configuration of the algorithms has to be added. The goal is to validate the functional correctness and completeness of the application as well as to prepare the application for the mapping onto the MP-SoC platform. The resulting application model is also referred to as the *Executable Specification* of the system.

Susbsequently the *system architecture* needs to be defined. This task covers not only the specification of the MP-SoC hardware, but also the mapping of the application model onto this hardware. The output of this task has a dominant impact on the resulting performance, efficiency and price of the complete system. For this reason, the specification of the system should be carried out as thorouh design space exploration to optimize the quality of results components needs to be defined. Questions to answer encompass the coarse grain separation of the application tasks into hardware and software, the number of processors, how different pieces of SW and algorithms will access main memory and which interconnect approach can deliver the required communication bandwidth. The main body of this book proposes a methodical approach to the design space exploration task.

Another design task within the MP-SoC desing phase is the *embedded SW development*. Here the parts of the application model to be executed in SW are refined to the final implementation. This covers for example the development of firmware and low level SW that needs to run on the hardware platform. This task is also related to the full MP-SoC plaform, because this kind of Hardware dependant Software needs to developed in the context of the actual platform.

Finally there is Hardware-Software co-verification task. During this task the hardware implementation code (RTL) is verified in the context of the platform. Vice-versa this is also the task to validate the correct operation of SW in the context of actual hardware.

All these tasks of the MP-SoC design phase will be referred to in the course of the discussion of state-of-the-art tools and methodologies for System Level Design in chapter 5. Especially the discussion of use-cases for Transaction Level Model in section 5.2 is tighlty linked with this classification of the design tasks.

4.3 Abstraction Mechanisms

In order to solve the problems of the different design phases, engineering of integrated circuits has always employed models on *different levels of abstraction*. A model serves as a unique, idealized description of the considered system, and the degree of abstraction characterizes the type of model used in the respective design phase. Goal of abstraction is to provide a description of the system, which is simple enough and yet sufficiently accurate to enable the necessary investigations, take design decisions and proceed to the next design phase. Indeed, the design-flow of an embedded system can be considered as a sequence of steps which successively reduce the degree of abstraction in the system model.

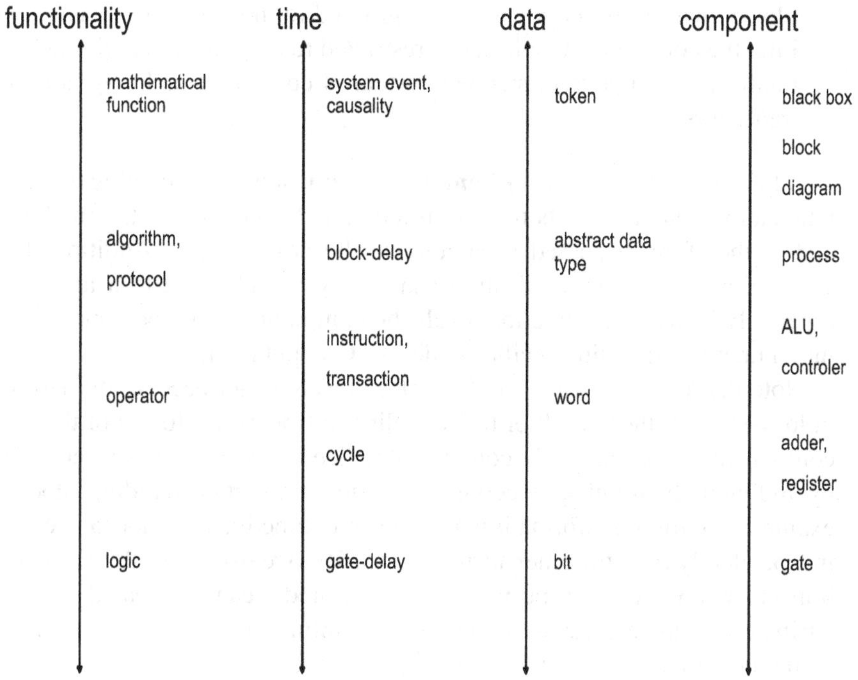

functionality: mathematical function; algorithm, protocol; operator; logic

time: system event, causality; block-delay; instruction, transaction; cycle; gate-delay

data: token; abstract data type; word; bit

component: black box; block diagram; process; ALU, controler; adder, register; gate

Figure 4.1. Modeling Attributes [84]

As depicted in figure 4.1, the modeling space can be classified into four orthogonal model attributes [84]:

- The **Functionality** refers to the modeling of the system behavior. On the highest level of abstraction, the functionality is condensed to pure mathematic expressions. Later the functionality is refined to operators, which in turn are finally mapped to logic gates.

- The **Timing** model captures the temporal properties of the system. Here the degree of abstraction ranges from causality of events to physical timing of transistors and wires.

- The **Data** representation in the model can also effectively subjected to abstraction in order to hide implementation details. Higher level data resolution is reduced to tokens and Abstract Data Types (ADT), whereas lower levels employ word or bit representations.

- The **Component** granularity describes the finest resolution of the sub-blocks, which are hierarchically composed to the complete system model. First the component resolution is restricted to coarse-grain building blocks, finally the complete embedded system is composed of fine-grain silicon transistors.

A further attribute would be **communication**, however modeling of the communication heavily affects the construction and execution of the model itself and is therefore discussed separately in the next section. Additionally, the abstraction of the communication can hardly be subjected to a linear scale, especially at higher abstraction levels the communication modeling bifurcates into numerous, domain specific Models of Computation.

Note that the refinement of abstraction levels is not necessarily performed in lock-step, neither for all of the modeling attributes nor for all of the system components. Instead, it is common practice to manage design complexity by individually refining a certain attribute or a certain building block. For example, the timing information of an abstract functional model can be refined independently from the other attributes in order to reason about the architectural impact. For verification purposes a hybrid model can be created, where the attributes of an individual component are refined to the synthesizable level while the remaining systems is kept on higher level of abstraction.

4.4 Models of Computation

The disciplined creation of a system model requires a modeling language and a well defined execution semantic coordinating the activation of the individual blocks. In that a Model of Computation (MoC) is composed of two parts:

- The *coordination language* describes basic execution semantics with respect to properties like parallelism, synchronism, reactivity and provides the abstracted communication mechanism.

- The *host language* provides the language elements for the specification of the system models.

Edwards and Sangiovanni-Vincentelli have conceived a widely adopted Tagged Signal Model as a formal framework to compare different MoCs [2] and provide a classification of popular MoCs [85]:

Timed Models of Computation. are characterized by the total temporal ordering of all occurring communication events. The most prominent example is the discrete event simulation MoC, which defines the execution semantics for HDL simulators.

Further examples of timed MoCs are synchronous languages like Esterel [86], Lustre [87], or Signal [88], where the events of all communication signals are constrained to occur at identical time stamps [89]. Thanks to their sound mathematical foundation, synchronous languages have gained adoption for the specification, analysis and code-generation of reactive control-dominated applications.

Untimed Models of Computation. are characterized by the fact, that communication events are only partially ordered. However, various untimed MoCs are popular for the specification of both data and control dominated applications:

- Data-Flow MoCs are heavily employed for algorithmic modeling and analysis of signal processing applications. The basic Kahn Process Network (KPN) [90] as well as a multitude of derived data-flow models like Synchronous Data Flow (SDF) [91], Integer-controlled Data Flow (IDF) [92] or hybrid models like the Process Coordination Calculus (PCC) [93] have been developed to optimize simulation speed, expressiveness, and/or analysis capabilities.
- Communicating Sequential Processes (CSP) [94] and Calculus for Communicating Systems (CCS) [95] are prominent untimed MoCs which are based on sequential processes that communicate using a rendezvous communication mechanism. Of course the events involved in the rendezvous are totally ordered and synchronous, however all other events are only partially ordered. For example the formal execution semantics of the standardized Specification and Description Language (SDL) [96] are defined using a CSP algebra based network of communicating processes [97]. SDL is widely used for the specification of communication protocols.

The definition of a proper MoC has long been considered to be the "silver bullet" for system level design and by that for the solving of the design productivity crisis: initially, the complete system functionality is to be created using the ideal MoC, which provides highest modeling efficiency, simulation speed, and

[2]Please refer to section 7.1 on page 79 for a more elaborated introduction of the Tagged Signal Model

smooth IP reuse. Next, the initial specification would be automatically verified using formal verification technology and metrics like determinism, causality, dead-lock absence, consistency, completeness, and fairness. The golden system specification would then provide the foundation for an automated path to design space exploration to take functional and architectural design decisions. Finally, system level synthesis would be applied to the partitioned system specification providing an automated path to implementation.

As further discussed in chapter 5, research on Models of Computation so far has failed to achieve any of the high aimed targets to support the MP-SoC platform phase. These efforts have produced a multitude of different *vertical* MoCs, which each address the abstract modeling of the data-plane processing for a particular application domain during the functional design phase. Within one vertical MoC the design methodology and tooling is well defined. However, the only well adapted *horizontal* MoC is the discrete-event model, which is employed for Hardware simulation.

4.5 Object versus Actor Oriented Design

This section briefly introduces the two major specification styles used for control- and data-plane processing, respectively. According to the multi-level SoC design classification in section 4.2, this specification phase is entirely situated in the functional phase.

Object Oriented Programming (OOP) is a powerful abstraction mechanism, where data and functionality is partitioned and encapsulated inside classes. OOP based languages like UML, C++, or Java are widely adopted in engineering of arbitrary SW and are rapidly gaining importance for the specification of embedded control-plane processing. OOP components interact primarily by sequentially transferring control through method calls. This sequential nature of OOP hinders the intuitive specification, analysis and refinement of the inherent parallel data-plane processing tasks.

For this purpose the actor-oriented abstraction scheme has been conceived [98], where parallel objects interact by sending and receiving messages. Within an actor-oriented design environment, the designer can focus on the specification and analysis of the algorithmic behavior of the individual tasks whereas the communication and synchronization aspects are handled by the underlying parallel Model of Computation [99].

Examples for actor-oriented design languages are the Ptolemy library [100], the Metropolis metamodel [101], SpecC [102] and SystemC 2.0 [18]. As a similarity, all of those share the notion of concurrent processes communicating through communication channels. Additionally, actor-based design languages achieve high modularity in communication modeling by using the Interface Method Call (IMC) principle [103], [104]. This is essential to support the complete variety of different MoCs.

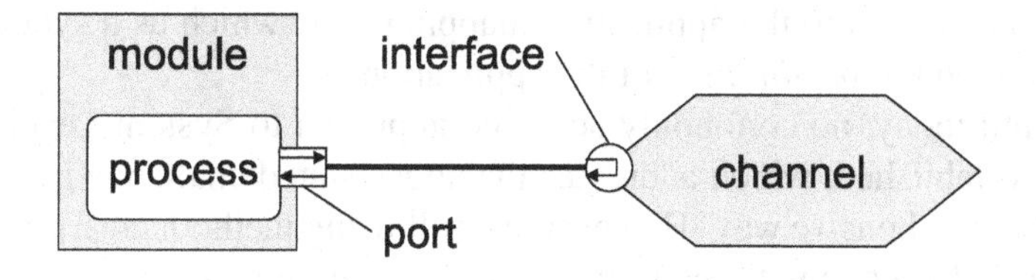

Figure 4.2. Interface Method Call Principle

As shown in figure 4.2, the IMC mechanism is realized by a set of language elements for modules, ports, interfaces and channels. Processes modeling the behavior are wrapped into modules and access communication services through ports. The available methods are declared in the interface specification and implemented by the channel. Thus the access methods in an interface reflect the specialized properties of the communication style implemented by an particular channel.

In this way actor-oriented design languages offers a *generic Model of Computation*, which in case of SystemC is based on an event driven simulation kernel [18]. Channels serve as containers for communication and synchronization. The user can extend the generic MoC by creating his own *methodology specific channel library* [105].

Summary. The design style for initial specification differs significantly for control- and data-plane processing portions of the application: Object Oriented Programming is widely employed to cope with the huge functional complexity of control-plane processing. Instead, the algorithmic complexity of data-plane processing is resolved by the actor-oriented programming style, which preserves the inherent task-level parallelism. The following section introduces modeling styles for the MP-SoC platform phase.

4.6 System Level Design Requirements

So far this chapter has introduced basic aspects for the design of complex Integrated Circuits before the implementation phase. In summary, a set of abstraction mechanisms with respect to the functionality, timing, data, component and especially communication has to be available to focus the designer on the respective design task.

In the context of a complete IC design flow, the MP-SoC platform design phase is all about defining a feasible and cost effective platform architecture as well as finding a spatial and temporal mapping of the application. As the particular challenge of System Level Design, the architecture definition and application mapping have to be considered jointly by taking the full functional and architectural complexity into account. In case of a fixed target platform,

SLD is reduced to the application mapping task, which as a synonym term is also called the *partitioning* of the application.

Until today, no commonly accepted approach to System Level Design has been established, which addresses the above stated challenge in a satisfactory and comprehensive way. However, the following methodical aspects have been identified to cope with requirements of System Level Design:

- **Orthogonalization of concerns** with respect to all modeling attributes [34] generally enables a divide-and-conquer approach to System Level Design.
- In particular, **Separation of interfaces and behavior** according to the interface based design paradigm motivated by Rowson [70] fosters successive communication and structural refinement as well as IP reuse.
- **High modeling efficiency and simulation speed** is mandatory to handle the high complexity of SoC designs.
- **Incorporation of hardware specific concepts** like timing, reactivity, parallelism, and determinism to express the impact of the platform architecture.
- **Incorporation of software specific concepts** like Object Oriented Programming, Operating System (OS) encapsulation, Inter Process Communication (IPC), process concurrency, as well as the creation, mutual preemption, and termination of processes to enable smooth integration of the embedded Software part.
- Support for **Verification and Validation** verification, to first gain evidence on the highest possible level of abstraction, that the correct system is being developed and all performance and cost requirements are met (validation). Later, the validated specification should be reused as a golden reference model for the subsequent refinement, IP integration and implementation steps (verification).
- **Seamless transition** between design phases and abstraction levels from system to gates to avoid long iteration cycles caused by gaps in the design flow.

Apart from these basic requirements, currently even the right design paradigm to cope with the emerging MP-SoC complexity crisis is yet to be agreed on [106]. Competing SLD paradigms and languages put different additional requirements like e.g. unified platform API smooth IP integration, tool automation, and/or intuitive user interaction into focus.

Chapter 5

RELATED WORK

This chapter introduces existing languages, design paradigms and tool environments for the MP-SoC platform design phase. The discussion is based on the methodology requirements stated in the previous section as well as the capability of the approach to cope with current and emerging architecture trends as summarized in section 3.4.

The multitude of related work in the area of System Level Design cannot be exhaustively covered in this book. Instead, the focus is on the identification of relevant concepts and a brief introduction of prominent representatives from the respective direction. Please refer to [107] for a more comprehensive overview.

At first, elementary techniques and early approaches to HW/SW Co-Design are introduced. The second part is reserved for an in-depth discussion of SystemC based Transaction Level Modeling (TLM), which has become a widely accepted and well supported abstraction layer above RTL. Afterwords, the subject turns to ongoing research projects addressing the emerging complexity crisis in MP-SoC design.

5.1 Traditional HW/SW Co-Design

The term HW/SW Co-design has been introduced in the early 1990s to describe the transition from board-level to chip-level integration of programmable microprocessors and ASICs [6].

Over the last decade, HW/SW Co-design has been focused on the development of elementary technologies like the definition of system level specification languages, HW/SW co-simulation and system synthesis algorithms.

The following sections review the results of these research efforts, which are of importance in the context of this book. A more complete survey is given in [6] and [85].

5.1.1 HW/SW Co-simulation

Early on, HW/SW Co-simulation has been recognized as a necessary ingredient for HW/SW Co-design. In a very general approach, the Ptolomy Co-simulation framework has been developed to investigate the mixing of different Models of Computation [100].

For the practical purpose of HW/SW interface verification prior to silicon fabrication, first HW/SW Co-simulation prototypes linked Hardware Description Language (HDL) simulators to an ISS executing the Software part [108]. Soon, HDL/ISS Co-simulation environments like Seamless CVE [109] became commercially available and are still widely employed. However, this HDL/ISS approach is severely limited by the slow simulation speed of the HDL simulator, especially in case of large systems with several ISSes and significant hardware portions [110].

To extend the use model of HW/SW Co-simulation towards performance simulation and development of Hardware dependent Software (HdS) [111], the concept of flexible hardware abstraction levels has been developed [112], where accuracy can be traded against simulation speed. Maximum simulation speed can be achieved by using compiled ISS technology [113] together with highly abstract functional SystemC models of the hardware part [114].

5.1.2 System Synthesis

The original goal of HW/SW Co-design was to reach the same degree of tool automation known from RTL synthesis, i.e. a formalized system specification is automatically partitioned and synthesized to the optimal target architecture.

Early attempts in this direction were Vulcan from Standford [115, 116], COSYMA from TU Braunschweig [117], and LYCOS from TU Denmark [118]. These projects are focused on the automated partitioning of a homogeneous, C-like input specification into dedicated hardware and software running on a single microprocessor. Subsequent projects like Cosmos from TIMA Laboratories [119], and SpecSyn from UC Irvine [120], extended the target architecture towards multiprocessor platforms.

However, automated HW/SW partitioning and System Synthesis have never gained industrial relevance. For one reason, the partitioning decision metric is restricted to worst case execution time, whereas other important metrics like average performance, cost, and power dissipation are not taken into account. Additionally, even the worst case execution time proved to be hard to estimate in the general case of parallel, data dependent, and interleaved software execution.

In a matter of fact, HW/SW partitioning and automated synthesis is still not recognized as a dominant issue, since the performance critical data-plane processing parts of the application are developed in a block oriented way. Instead, system architects are interested in the impact on performance of a specific target architecture. To partly automate this mapping, *Communication Synthesis* and *HW/SW Interface Synthesis* emerged as new branches of HW/SW Co-design.

5.1.3 Communication Analysis and Synthesis

As already discussed in section 5.1.1, the required modeling effort as well as the achievable speed of conventional HDL based architecture simulation is not sufficient for design space exploration of complex communication architectures. This has triggered the development of techniques for the analysis of communication requirements and synthesis of the communication architecture.

Several academic prototypes for static communication analysis and synthesis have been developed [121–123] or Communication Synthesis became the focus of general HW/SW Co-design frameworks [124, 125]. However, Communication Synthesis faces problems similar to general System Synthesis: communication time estimation is based on the restrictive assumption, that processing and communication are statically scheduled. Additionally, quality of results is afflicted by either overly optimistic [123, 125] or worst-case [122] communication contention scenarios.

In a hybrid approach to communication analysis, Lahiri et al. combined the accuracy of simulation and the speed of static estimation [126]. In a first step, this approach creates an event graph from an initial HW/SW Co-simulation run. Based on this accurate traffic profile, the subsequent performance estimator can effectively analyze the impact of bus parameters like e.g. width, priorities, protocol overhead. This approach achieves sufficiently accurate results for fast design space exploration and supports various bus features like bridged segments [127] or TDMA arbitration [128].

In general Communication Analysis and Synthesis techniques need further advancement to cope with emerging Network-on-Chip architectures. First steps into this direction are currently undertaken. A straight forward attempt is to instantiate the NoC library elements (routers, network interfaces, links) from a high-level view of the SoC floorplan [129]. The selection of the actual library elements can be supported in different ways: In a application-centric approach, the network topology can be generated from a communication graph of the application [130]. As an alternative architecture-centric approach, the communication architecture can be refined from an abstract channel view via a network topology view towards a micro-architecture view [131].

So far the analysis of Network on Chip architectures is performed using handcrafted simulation models, which are mostly based on SystemC [132, 133, 129]. The absence of standardized APIs, abstraction levels and modeling frameworks beyond the plain SystemC language so far hinders the creation of interoperable IP models for NoC architectures. Some of the current projects working on a unified modeling environment for the exploration of NoC architectures are discussed in section 5.3.3 below.

5.1.4 Interface Synthesis

As the basic idea of HW/SW Interface Synthesis, the designer itself decides on the partitioning and architecture mapping but the *realization* of these decisions are supported by automating the tedious task of generating the required Software driver functions as well as the Hardware glue-logic.

CoWare commercialized the Interface Synthesis technology developed at IMEC [134]. The proprietary CoWareC input specification is based on the sequential Remote Procedure Call (RPC) communication paradigm, where functional tasks communicate through blocking master-slave channels. Based on the CoWareC system model and the user defined mapping, the hardware and software communication interfaces are generated automatically. The mapping decisions are verified using the accompanying Co-simulation framework with support for ISS, HDL and CoWareC integration [135]. With this *Napkin-to-to-Chip (N2C)* framework CoWare became the dominant ESL tool provider. Recently the technology has been ported to the SystemC based CoWare Platform Architect product line [19].

5.2 SystemC based Transaction Level Modeling

The MP-SoC platform phase is concerned with the system architecture specification as well as the application mapping. Therefore, abstraction concepts on this level have to support the joint consideration of application and architecture. On the other hand, the high level of detail inherent to Register Transfer Level (RTL) implementation models prohibits the investigation and optimization across heterogeneous communication and processing elements.

Significant research has been spent on the definition of the appropriate System Level Design language. Today SystemC is generally considered as the standard language for all kinds of SLD tasks.

5.2.1 History of SystemC

SystemC has initially been conceived to replace VHDL and Verilog as a Hardware Description Language [136]. For this reason it naturally provides all hardware specific concepts e.g., time, parallelism, and hierarchy. In 1999 the Open SystemC Initiative (OSCI) [137] has been founded by a consortium of EDA tool companies, semiconductor houses and IP providers to establish SystemC as an industry standard. With version 2.0 SystemC has been thoroughly revised to become a fully elaborated actor oriented design language [103]. The incorporated Interface Method Call (IMC) principle enables a clean separation of interfaces and behavior as well as orthogonalization of further modeling attributes. Additionally, all kinds of methodology and application domain specific Models of Computation (MoC) can be implemented on top of the generic event-driven SystemC simulator [18]. By that SystemC 2.0 enables a smooth transition from functional phase to the MP-SoC platform phase, e.g. hybrid simulation of an architecture model in the context of an algorithmic Data-Flow model [138].

Since SystemC is a native C++ library, it inherently supports Object Oriented Programming. However further software concepts like dynamic process creation and termination as well as user defined process scheduling is not supported until the availability of release 3.0 [139].

Recently the final version 2.1 of the language has become an official IEEE standard [140]. OSCI continues to foster the SystemC ecosystem of tools and IP, as well as to incubate the development of additional libraries and methodologies. Examples for these incubation activities are the working groups for the development of SystemC Verification (SCV) library [141], for the development of the Transaction Level Modeling (TLM) kit [142], and for the definition of the synthesizable subset of SystemC. Potentially these developments are incorporated into the SystemC language itself.

5.2.2 History of Transaction Level Modeling (TLM)

The characteristic property of TLM is that the pin-level communication interface of RTL models is replaced by a set of interface methods. In theory, this IMC based communication mechanism is provided by all actor-oriented specification languages. However, only SystemC is gaining industrial acceptance and commercial tool support for TLM based system architecture modeling and can henceforth be considered as the driver language for the further development and deployment of the TLM paradigm.

The TLM Standard

Early work on SystemC based TLM has demonstrated the potential in terms of increased simulation speed and modeling efficiency [18]. As the result of this success the TLM working group has been established to work on a standard interface for SystemC based TLM. The resulting TLM 1.0 API provides a concise set of interfaces with well defined semantics [142].

The basic TLM API consist of a bidirectional *transport* and a set of unidirectional *put* and *get* interfaces. The bidirectional transport has *blocking* synchronization, i.e. implementation of the interface is allowed to call `wait(.)`. The unidirectional interfaces are available in a *blocking* and a *non-blocking* version.

These interfaces can be seen a foundation layer for the creation of more advanced TLM interfaces, which serve a specific methodology or model a specific communication protocol. Please refer to appendix A for an introduction of the TLM 1.0 API.

The TLM Abstraction Levels

Apart from the definition of the foundation API, the OSCI TLM working group has also worked on the definition of a common set of abstraction levels [143, 144]. Although this work is not finished, this paragraph gives an overview of the current proposals.

The two cycle-level TLM layers **Bus Accurate (BA)** as well as **Cycle Callable (CC)** are generally agreed on and are already well supported by state-of-the-art ESL tools [19, 20] and IP models [145]. These levels are particularly suitable to create a cycle-accurate prototype of the system architecture, where the (usually cycle-accurate) Instruction Set Simulators (ISS) of the programmable architectures are connected to cycle- and bit-accurate models of memories, communication resources and peripherals [146]. In comparison with RTL, the simulation speed of cycle-level TLM is improved by two orders of magnitude without sacrificing timing accuracy. BA and CC abstraction levels are distinguished by the fact, that BA captures a transaction within a single method call, whereas CC models provide separate methods for every phase of a transaction.

The **Programmers View (PV)** abstraction levels address early integration of (usually instruction accurate) ISSes for SW development purposes. PV provides a bit and address-map accurate view of the MP-SoC architecture context for the programmable processing elements. PV is based on the bidirectional blocking transport API of the OSCI standard, i.e. represent a purely sequential Remote Procedure Call (RPC) modeling paradigm similar to the CoWareC language [134]. Like this the PV calls from an ISS to a peripheral component do not return before the peripheral component has executed the triggered functionality. This simple synchronization scheme minimizes the interaction with the SystemC kernel and therefore yields high simulation speed.

The PV abstraction level is well defined and already widely used throughout the industry to enable an early start of the SW development on the SoC target platform [147]. There is however currently no commonly agreed concept adding timing annotations. The **Programmers View with Timing (PVT)** is still under discussion in the TLM working group.

Communicating Processes (CP) and **Communicating Processes with Timing (CPT)** have been discussed as suitable abstraction levels for the generic architectural modeling and application mapping of parallel data-plane processing tasks. However, CP and CPT are currently even further away from standardization than PVT. Instead, the aspect of generic architectural modeling has been picked up and standardized by the System Level Design working group of OCPIP.

5.2.3 The OCPIP Channel Library

Early attempts to unify communication interfaces at the architectural level like VSIA [148] have not been widely adopted. The Open Core Protocol International Partnership (OCP-IP) [22] is getting a lot of traction throughout the industry. OCPIP provides a high configurable SoC protocol and their System Level Design working group has worked from the early days on Transaction Level Modeling [149].

OCP API		Data Accuracy	Timing Accuracy	Addressed Design Problems
TLM	TL3	packets	total event ordering, burst-level annotation	functional specification, generic architecture exploration
	TL2	burst of words	burst-level and word-level annotation	OCP architecture exploration
	TL1	word	cycle accurate	100% cycle accurate performance profiling

Figure 5.1. TLM Abstraction Levels as defined by OCPIP

An overview of the abstraction level as defined by OCPIP is depicted in figure 5.1 [149, 150].

- The lowest level is called Transaction Layer 1 (TL1) and provides a fully cycle accurate model of the OCP protocol. This level is fully aligned with the CC abstraction level from OSCI.
- The next higher level is called TL2 and represents basically a cycle-approximate abstraction of the OCP protocol. The API is still pretty rich and contains a large number of OCP specific features like e.g. thread-busy, handshake-timing, or sideband signals. The timing is not cycle accurate, but can be annotated to a near-cycle accurate level (see section 4.7 of [151]).
- The Transaction Layer 3 represents the highest abstraction level of the OCP standard and can be considered as the *protocol agnostic subset* of TL2. The TL3 API is limited to a concise set of primitives, which are essential to model timing approximate on-chip communication. The TL3 channel supports a dual-parameter timing model (please refer to section 6.2.1).

The early implementation of the TL3 API was based on the generic channel, which is no longer supported by OCP-IP. The latest OCP-IP TLM package (v2.1.2,[151]) does include the revised TL3 API, which is now implemented in top of the OSCI TLM API (please refer to appendix B for more information).

5.2.4 TLM Summary

PV TLM platforms for early SW development as well as cycle-level TLM for HW/SW and TLM/RTL co-verification are successfully deployed throughout the industry [152]. However, both use-cases solve only parts of the challenges during the MP-SoC design phase. Especially the architecture definition and task partitioning is not adequately addressed:

PV platforms simulate very fast and are well suited for SW development. Unfortunately they do not contain sufficient timing information for architectural investigations. The blocking semantics of the underlying bidirectional transport API hinders the smooth annotation of further timing information.

Cycle accurate models of the SoC platform are too detailed and too slow for architecture definition and task partitioning. First, the effort to create such a cycle-accurate model of the complete platform is way too high to allow for the investigation of a large number of architecture and application mapping alternatives. Second the reachable simulation speed in the order of 100k cycles per second is not sufficient for the analysis of large design parameter choices.

As a result, the exploration of broad design spaces is still a cumbersome process in cycle-level TLM based design flows: cycle-level TLM communication models have architecture specific interfaces. Thus, every time the designer is inclined to explore a new communication architecture he has to change the interface of the connected functional models.

For this reason the Design Space Exploration framework described in this book deploys a generic synchronization interface, which provides the same primitives as the newly standardized OCP TL3 API. Obviously, the TL3 API presents the best fit for this purpose. It is compliant with the OSCI TLM standard. Additionally, it is of reasonable complexity, and yet offers sufficient expressiveness to meet the accuracy requirements for design space exploration.

By deploying SystemC based Transaction Level Modeling the framework is nicely integrated into the flourishing ESL ecosystem. Like this the work presented in this book is interoperable with the PV and cycle-accurate modeling methodologies and can benefit from the commercial tool support, available IP models, and established ESL design methodologies [153].

5.3 Current Research on MP-SoC Design Methodologies

HW/SW Co-design research has brought forth several core technologies like HW/SW Co-simulation and Interface Synthesis, which are widely employed and well supported by mature commercial tools. As a further success, SystemC is considered as the emerging standard language for System Level Design and the cycle-level Transaction Level Modeling paradigm has become an accepted and well supported abstraction layer above RTL.

The rather utopian System Synthesis scheme is replaced with more pragmatic approaches, which keep the designer in the optimization loop. Current directions can be coarsely separated into a IP centric *bottom-up* Component Based Design paradigm, the Network-on-Chip driven Communication Based Design and the *top-down* exploration and refinement environment.

5.3.1 Component Based Design

Component Based Design [154] is founded on the assumption, that the processing elements and communication templates are available IP blocks. Hence, the design flow of MP-SoC platforms can be seen as bottom-up composition of parameterizable IP library elements.

A prominent representative of this paradigm is the ROSES design framework from Jerrahya et al. [155, 156]. This work is founded on a library of processors and parameterizable communication templates, and employs protocol synthesis technology for automatic generation of the required hardware wrappers [157, 158].

In opposition to traditional HW/SW Interface Synthesis, the Software drivers are not synthesized. Instead, the IP library also contains a stack of SW abstraction levels encapsulating OS and hardware driver services [159], which are consistent with the on-chip communication templates [160]. By that, the embedded SW is developed on top of a set of well defined Service Access Points (SAP) [161], which are completely independent from the hardware context.

Undoubted, this approach speeds up the development of the embedded control-plane processing, where the processing elements are limited to a small selection of general purpose micro-controllers. On the other hand, the performance loss due to the SW abstraction stack, wrapper generation and generic communication templates is prohibitive for data-plane processing. Additionally, the a posteriori ISS/HDL based performance simulation is way too slow for MP-SoC design space exploration.

In a complementary data-plane centric Component Based Design approach, Philips Research has developed the Eclipse architecture template and corresponding design flow for multimedia processing [76, 162, 15]. The design flow is assumed to start with an initial algorithmic exploration using the Kahn Process Network (KPN) formalism. The architecture template is furnished to preserve the KPN Task Level Parallelism and communication semantics:

- The high performance signal processing tasks are mapped to application specific processing elements to execute.

- The KPN FIFO channels are mapped to either dedicated HW FIFOs [76] or memory regions [162].
- The Kahn process coordination is implemented by synchronization shells, which guard the execution of the processing elements.

The Eclipse architecture template improves the utilization of the Processing Elements by extending the synchronization shell with hardware support for task interleaving [15]. By that, the PEs can efficiently switch to another task in case the current task is stalled due to memory contention.

Recently the KPN formalism has been generalized to a Task Transaction Level (TTL) interface. This API provides more flexibility in terms of blocking and non-blocking semantics and is therefore applicable to wider range of control-oriented applications [163, 164]. This approach is still very much focused on the signal-processing and multi-media domains. This might change in the future, as a recent experiment has demonstrated the feasibility to map the TTL API on top of a general purpose RTOS [165].

Despite the effectiveness of this kind of domain specific frameworks, the tight formalism of the input specification and the restricted flexibility of the architecture template limit the eligibility beyond the considered application domain.

5.3.2 Communication Based Design

The Communication Based Design paradigm [66] also envisions MP-SoC platform design as a composition of reusable IP blocks. Different from Component Based Design, it omits the consideration of processing elements and is exclusively focused on the the conceptualization and implementation of the communication architecture. By that, Communication Based Design can be seen as the corresponding design paradigm to match emerging NoC architectures.

Inspired by earlier work on communication analysis, the *EDA centric* approach aims at the automated selection and configuration of the communication infrastructure from a comprehensive IP library. Along these lines, the recent work from UC Berkeley proposes constraint-driven communication synthesis [166, 167]. Driven by a constraint graph, the communication requirements are analyzed and an optimal communication architecture is selected and configured.

The *NetChip* project from Stanford university and Bologna university constitutes an other prominent EDA centric approach to communication based design [168]. This approach is also based on a library of scalable network components called *xpipes* [129], which are automatically instantiated and configured by an exploration tool named SUNMAP [130].

On the other hand, the *architecture centric* approach [169] assumes, that the NoC capabilities can handle arbitrary traffic conditions. By that the MP-SoC platform design is upfront bound to a fixed communication infrastructure, which merely needs to be configured to meet the communication requirements of the considered application.

However, both approaches presume a perfectly known communication pattern, where the application tasks are already partitioned and mapped to the respective processing elements. Thus, Communication Based Design can be considered as backend of a preceding design space exploration and application mapping phase.

5.3.3 Design Space Exploration (DSE) Environment

The goal of this approach is to take early design decisions with respect to system architecture and application mapping on the basis of an abstract performance model. For this purpose, the embedded application needs to be modeled together with the MP-SoC architecture at a high level of abstraction. Prior to the discussion of related work in this area, a brief introduction of common DSE principles is given. A thorough introduction of the topic including a survey of related work can be found in [170].

In concordance with the general Orthogonalization of Concerns requirement, the **y-chart** scheme [171, 172] emphasizes the separation of architecture and application models. Here the application models are associated with the target architecture in an explicit mapping step. The results of the performance simulation guide the system architect in taking optimal design decisions. By keeping the architecture models separated from the application, various architectural alternatives can be rapidly explored without touching the already validated functional application models. Additionally, both models are highly reusable across multiple design projects.

As another basic principle of design space exploration, the **abstraction pyramid** [173] emphasizes the interdependence of model abstraction, evaluation cost in terms of modeling effort and simulation speed as well as impact of design decisions. Hence, the initial DSE model has to be sufficiently abstract to allow for broad investigation of design alternatives. Later on, the model is successively refined towards lower abstraction levels and increased accuracy to verify early design decisions.

Apart from these common DSE principles, the various research projects discussed in the following paragraphs differ significantly in terms of application domain, abstraction levels and refinement mechanisms.

SPADE As an early attempt in creating a DSE framework for the signal processing application, the SPADE (System-level Performance Analysis and Design-space Exploration) project [174] first conceived the y-chart scheme and the abstraction pyramid as basic principles for interactive design space exploration. Typical for signal processing applications, the functional models are created according to the Kahn Process Network (KPN) formalism. For this purpose, SPADE provides the KPN communication primitives by means of a programming interface called YAPI [175]. After the KPN based algorithmic exploration is finished, the application model is bound to an architecture model by means of trace-driven co-simulation [176]. The mapping of the functional Kahn processes to processing elements and the Kahn FIFO channels to communication resources reveals the resulting performance through system simulation [173].

ARTEMIS The work of SPADE is continued by the ARTEMIS (Architectures and Methods for Embedded Media Systems) project [177], which is further divided into two branches. The Sesame project [178] improves simulation accuracy by refining the coarse-grain KPN application traces to fine-grain architecture models using Integer-controlled Data-Flow (IDF) based communication refinement technique [179]. Additionally, the risk of deadlock situations when mapping unbound Kahn FIFO channels to limited communication resources is removed by the introduction of an additional synchronization layer. More recently, this approach has been extended to automated DSE based on genetic optimization algorithms [180].

The Archer project branch takes a complementary approach to improve the mapping accuracy. Here the YAPI application model is first translated into a symbolic Control-Data-Flow-Graph (CDFG), which is closer to the final implementation [181]. Although the simulation falls behind trace-driven simulation, the accuracy is sufficiently high to investigate the benefits of Instruction Level Parallelism provided by the target architecture [182].

Due to the Kahn specification formalism, the Artemis project is still limited to dataflow applications.

The **Performance Network Approach** from Thiele et al. provides a more formalized mechanism for Design Space Exploration [183]. Here the application workload is represented as a set of abstract arrival curves and the system resources are represented as service curves. Like this the expected performance of the embedded system can be computed rather than simulated. Despite the very high abstraction level, this approach has been successfully applied to a real-life example from the networking application domain [184]

POLIS As another important pioneer of DSE, the POLIS approach particularly addressed the design and verification of control dominated reactive Systems [172]. The input specification is based on Co-design Finite State Machines (CFSM), which thanks to their formal semantics enable automated synthesis and verification. Especially for software design, the CFSM specification effort hinders general acceptance and limits the applicability to control dominated systems.

VCC In an early attempt to commercialize tooling for MP-SoC integration and application mapping, Cadence Design Systems developed Virtual Component Co-design (VCC) [185]. Despite incorporation of several innovative concepts like abstract architecture modeling [186] and virtual instruction set simulation [187]. However, neither SystemC (as a common SLD language) nor cycle-level TLM (as an established modeling paradigm above RTL) were available at that time. Altogether, the proprietary modeling language and methodology, the lack of IP support packages, the missing path to implementation as well as the complex user interaction hindered the acceptance of VCC beyond experimental evaluation projects [188, 189].

Metropolis The research work of POLIS is continued in the Metropolis project [101], which still emphasizes the value of formal semantics. The input specification is no longer restricted to CFSMs, instead the Metropolis framework is based on an actor-oriented design language called generic meta-model [190]. In comparison to less formalized actor-oriented design languages like SystemC and SpecC, the Metropolis meta-model comprises additional formal elements for the specification of functional, temporal and cost properties. As a major advantage, this enables the compositional verification of arbitrary design properties [191]. Additionally, the flexible meta-model enables the definition of a software friendly MoC for the actor oriented specification of performance relevant software portions [192].

Altogether, the Metropolis framework can be considered as competitive approach to the less formalized SystemC language. In principle, the development of methodology specific communication libraries like the work described in this book can be implemented on top of either SystemC or Metropolis. On the one hand side, the flexibility of SystemC is clearly advantageous for Object Orient Software Programming and the creation of fast and cycle accurate TLM platform models. On the other hand, the evolution of system level verification will decide, whether informal SystemC enhanced with formalized assertions will prevail over formal specification languages like the Metropolis meta-model.

SpecC [102] is a System Level Design language quite similar to SystemC 2.0 [104]. As a major difference to the native C++ library approach of SystemC, SpecC is a subset of the ANSI C language augmented with a set of keywords for architecture modeling. The resulting benefit is a more concise and formalized modeling style with clear defined semantics for execution [193] and tool supported refinement transformations [194]. On the other hand, this strength is paid with a lack of flexility, which is necessary to tailor modeling and tooling for individual requirements.

Recently, increased support for Software specific concepts like RTOS modeling has been supported [195, 196], but the lack of Object Oriented Programming concepts still hinders smooth integration of Software development flows.

The Modeling Environment for Software and Hardware (**MESH**) project is concerned with modeling of heterogeneous MP-SoC platforms above the cycle-level ISS / TLM abstraction layer [197]. MESH perceives the MP-SoC mapping phase as the sequencing of logical events generated by functional processes into physical events. The concept of *Frequency Interleaving* is introduced to describe the event sequencing as an interleaved scheduling of the functional processes onto physical processing and communication resources [198]. By that, MESH inherently supports the high-level performance investigation of multi-threaded processing and communication resources [199]. In fact, schedulers are considered as the central modeling element to capture the dynamic and data-dependent nature of MP-SoC platform mapping [200].

Although the MESH project shares basic concepts with the work described in this book, the realization in terms of design methodology and tooling seems to be in a rather early stage. The modeling is based on proprietary language and simulator, so the MESH investigation is not integrated into any existing design flow [201]. So far, the focus is on modeling of multi-threaded processing elements and no results on modeling complex Network-on-Chip based communication architectures has been reported. Recently this work has been focused on defining benchmarks for MP-SoC platforms [202].

MultiFlex is a research project at ST Microelectronics investigating SystemC based tooling and design methodology for heterogeneous multi-processor platforms [28]. Similar to commercially available SystemC based SLD frameworks, the platform modeling aspect is focused on cycle level TLM, where the embedded software is already executed on the target ISS. The StepNP platform modeling framework is particularly focused on networking applications [203] in that it supports the integration of application specific and hardware multi-threaded processing elements as well as Network-on-Chip based communication architectures.

Recently the MultiFlex project has focused on the definition of multi-processor programming models by applying Symmetrical Multi-Processing (SMP) and client-server type of distributed component object models to the embedded systems domain [204, 205]. The corresponding support in the MultiFlex toolset enables the efficient implementation of applications using these programming models on multi-processor platforms.

Although StepNP has conceived a mechanism to explore the task mapping of multi-threaded processing element at the ISS level, the investigation of the spacial and temporal task mapping requires significant effort for setting up software implementation related details like e.g. RTOS configuration, memory map, interrupt handling etc.

The On-Chip Communication Network (**OCCN**) is another project in ST, which develops a SystemC based framework for modeling, simulation and design space exploration of complex communication architectures [206, 207]. OCCN advocates a protocol refinement methodology, which follows the layered ISO/OSI model [208]. Published documentation and code examples focus on the investigation of higher layer communication protocol aspects, like e.g. packetization (including header parsing, packet classification, lookup, data encoding, and compression), packet admission control, or congestion avoidance.

GreenBus is the project name for a collection of work aimed at providing an open source modeling framework that will enable system-on-chip (SoC) designers to exploit SystemC modeling techniques easily and efficiently early in the design cycle [209]. The emphasis of the project is on model interoperability and the results have been submitted to the Open Source SystemC Initiative Working Group on Transaction Level Modeling (OSCI TLMWG). GreenBus provides a SystemC 2.1 style port-to-port bound bus fabric which could be configurable (using SPIRIT compliant XML) to represent any bus (at a Programmers View, cycle-accurate, and a cycle-count approximate level of abstraction). It comes complete with a native ability to have user APIs such that a user can choose their interface independent of the bus fabric itself.

MPArm is an open research project coordinated by University Bologna [210]. Similar to the STepNP platform modeling framework of MultiFlex, the MPArm platform ingrates different processor and interconnect IP at the cycle-accurate level as well as Software stacks to investigate architectural alternatives in Multi-Processor platforms.

The **ARTS** abstract system-level modeling framework from TU Denmark also advocates the proposition of this book, in that resource sharing of processing elements and interconnect nodes must be jointly investigated to explore the complete design space [211]. The ARTS framework from Madsen et al. takes a Software-centric view onto the SoC platform [212] and models architectural elements at the level of arbitrated resources [213, 214].

The basic concept of an C++ based reactive process network exchanging Abstract Data Types has been investigated by Post et al. [215, 216] in the context of the **GRACE++** project. Significant effort has been spent on the development of techniques for efficient object oriented application modeling and seamless co-verification of RTL implementation models. The resulting GRACE++ methodology and simulator library distinctively supports the modeling and verification of complex telecom applications. This approach has been successfully applied to the design and verification of an ATM switch system [217].

The work described in this book is founded on the early work on GRACE++ and still profits from the OOP modeling techniques and RTL co-verification flow. On the other hand, the GRACE++ coordination library has been ported to SystemC and fundamentally revised to be applicable to arbitrary application domains. Additionally, the scope of covered architecture models has been extended from pure hardware blocks using point-to-point communication towards arbitrary processing elements and the full scope of on-chip communication architectures.

5.4 Summary

To summarize the current directions in MP-SoC platform design, one can observe a huge variety of proposed solutions. Different approaches emphasize different aspects as IP re-use, formal verification, or design space exploration. However, there is currently no comprehensive disciplined approach addressing the full design complexity of emerging Network-on-Chip enabled Multi-Threaded/Multi-Processor SoC platforms.

Cycle-level TLM has matured as virtual prototyping technology above RTL, but is not sufficiently abstract to achieve sufficient modeling efficiency and simulation speed. So far, there is no established approach at hand to capture and evaluate in an efficient way all aspects of heterogeneous MP-SoC platforms and application mapping.

Before a certain degree of maturation in SLD modeling above cycle-level TLM is reached, the Design Space Exploration environment is the most promising approach to drive the evolution of a common MP-SoC design methodology. The evolution of cycle-level TLM and RTL shows, that in order to achieve broad support in terms of tooling and IP availability, first the right level of abstraction for the addressed design problem has to be identified. Second, a concise modeling style has to be defined, which takes full advantage of the abstraction level to achieve high simulation speed and modeling efficiency. Last, techniques for simulation acceleration and further tool automation of the established modeling style needs to be developed.

Chapter 6

METHODOLOGY OVERVIEW

The discussion now turns to the Design Space Exploration (DSE) environment, which has been developed in the course of the studies described in this book. Ultimate goal is to meet the System Level Design requirements as specified in section 4.6 and to cope with the full architectural complexity of emerging MP-SoC architectures as exposed in chapter 3.

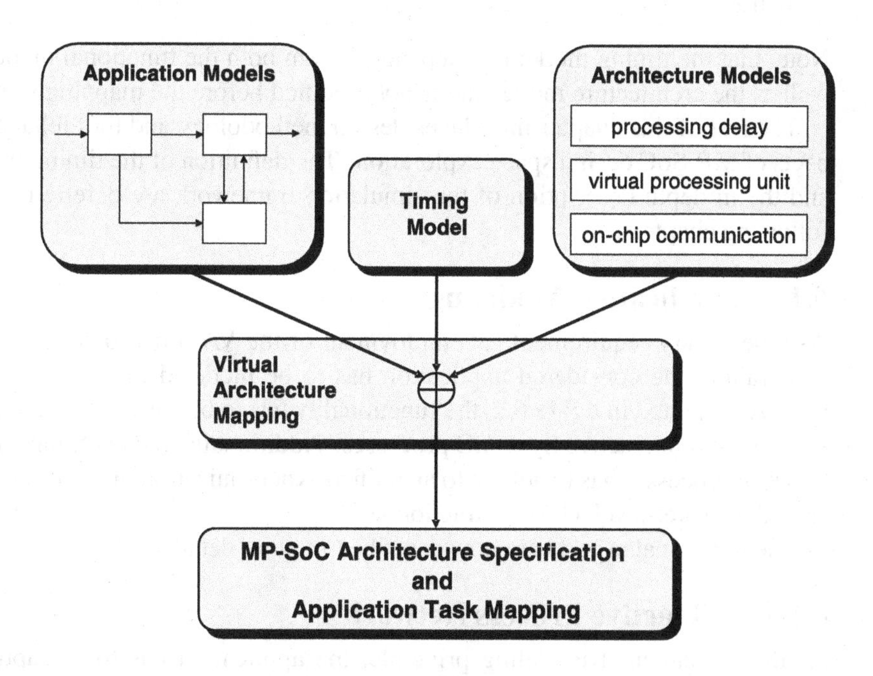

Figure 6.1. Virtual Architecture Mapping

As depicted in figure 6.1, the MP-SoC Framework follows the y-chart principle described before, where a set of functional application models is merged with a set of architecture models in a dedicated mapping step. In reference to the flexible and highly abstracted mapping mechanism, the developed embodiment of the y-chart principle is called *Virtual Architecture Mapping (VAM)*. Distinctively, Virtual Architecture Mapping comprises the following fundamental elements:

- A well defined abstraction level above cycle-level TLM for efficient modeling of embedded applications.
- A set of generic, parameterizable architecture models, which capture the notion of shared and resource limited architectural fabrics for communication and computation.
- A rigorous definition of a timing model, that embodies the performance of a selected application-architecture-mapping.
- An MP-SoC simulation framework featuring a declarative mapping mechanism to minimize turn-around times during the iterative architecture exploration cycle.
- A comprehensive set of analysis tools for functional and performance validation.

Note, that the timing model is independent from both the functional model as well as the architecture model and is not specified before the mapping step.

The rest of this chapter introduces design methodology and tool related aspects of MP-SoC design space exploration. The definition of the timing model and the in-depth description of the simulation framework are deferred to the following chapters.

6.1 Application Modeling

As a necessary requirement for employment of the VAM based design space exploration, the considered application has to be modeled in a well-defined way. As depicted in figure 6.2, the functionality has to be partitioned into a set of coarse-grain reactive SystemC processes. Additionally, the communication between processes has to adhere to an unified synchronization interface as well as to the packet-level TLM abstraction layer.

The individual aspects are now described in more detail.

6.1.1 Reactive Process Network

As the fundamental modeling principle, the application has to be captured as a SystemC based Reactive Process Network (RPN). The formal definition

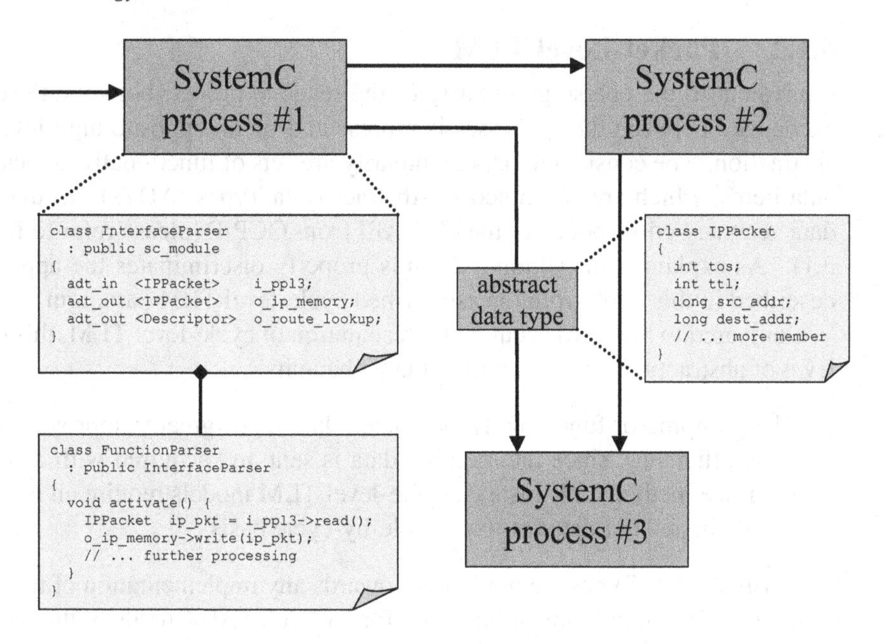

Figure 6.2. ADT Data Exchange

of the underlaying MoC is given in section 7.2, but intuitively this approach provides the following benefits:

- The actor-oriented paradigm naturally fits into every data-plane centric application development flow. Preserving the inherent task level parallelism, the block oriented algorithm specification can be converted into a functionally equivalent reactive process network.

- Since SystemC is a native C++ library, this approach also enables smooth integration of control-plane centric application parts. In particular, OOP based sequential functionality is wrapped into a reactive SystemC process. As demonstrated during numerous experiments, this effort remains reasonable thanks to the course granularity of the process network [215, 218, 219].

- The moderate number of coarse-grain reactive SystemC processes minimizes the number of process activations during the simulation. Compared to cycle-driven process activation, this greatly improves the simulation speed.

So to speak, a coarse-grain reactive process network represents a good compromise between the RPC-like sequential communication mechanism for control-plane processing and all the domain specific MoCs for data-plane processing. By that, the expressiveness of RPNs covers the complete range of embedded applications.

6.1.2 Packet-Level TLM

Equivalent to the coarse granularity of the reactive process network, the data exchanged between the processes is represented on an apposite high level of abstraction. The considered data granularity are sets of functionally associated data items, which are combined to Abstract Data Types (ADTs). In that the data accuracy corresponds to the TL3 API from OCPIP (please refer to figure 5.1). As explained in section 5.2.4 this property discriminates the approach described in this book from the established cycle-level TLM paradigm.

Compared to the word accurate representation of cycle-level TLM, this high level of abstraction offers a set of valuable benefits:

- The grouping of functionally associated data fields greatly improves modeling efficiency, since the required data is sent and acquired with a single interface method call. Instead, cycle-level TLM models require an FSM to stream in and out data sets on a cycle-by-cycle basis.
- Abstract Data Types are not biased towards any implementation of the data representation in terms of bit-width for particular ADT fields or the overall arrangement of the frame layout.
- The coarse granularity of communication further reduces the number of events during the simulation of the reactive process network. This immediately leads to increased simulation speed.
- In analogy with traditional network design environments like e.g. Opnet [220], the emerging Network-on-Chip era advocates the investigation of the on-chip communication infrastructure at the packet level.

Additionally, ADT based communication intuitively supports modeling of both abstract SW-SW communication principles like message passing as well as more HW oriented memory mapped communication.

In combination with the communication mechanism and the corresponding timing model, the packet-level TLM paradigm introduces a well defined level of abstraction into the design flow, which enables efficient functional specification and architecture exploration of complex MP-SoC platforms.

6.1.3 Generic Synchronization Interface

According to the IMC based communication scheme[1], the elementary task of a synchronization interface is the specification of the set of methods. These methods have to be implemented in the connected channel and represent the communication services available to the process. The careful definition of a

[1]Please refer to section 4.5 on page 40

unified synchronization interface is of threefold importance for the virtual architecture mapping based design space exploration.

First, a versatile y-chart based design space exploration environment calls for the capability to arbitrarily mix and match application and architecture models. Hence, the offered synchronization services have to be completely orthogonal from any architectural component for communication and computation. For example, the designer can decide to change the communication architecture from point-to-point towards a bus centric architecture modifying neither the synchronization interface nor any of the participating functional modules. Advanced communication services as offered by NoC architectures [8] are of course not covered. Instead, these services require a communication protocol on top of the basic synchronization interface.

Second, the timing annotation mechanism is integral part of the mapping step and therefore incorporated into the synchronization interface. In adherence to the orthogonalization of concerns paradigm [34] the functional models are not obscured with performance related timing annotation. On the other hand, accuracy is a special challenge for timing annotations in a packet-level TLM environment. The synchronization interface has been furnished to capture the observable events at the begin and the end of a packet transfer. This enables the creation of near cycle accurate communication models despite the coarse granularity of the data transfer [23].

Third, an protocol agnostic synchronization interface fulfills the interface based design principle [70], in that refined communication and processing models can be integrated into the abstract context by the use of adapters. This enables successive refinement and component-wise verification of refined models against their abstract specification model.

As a specific property of packet-level data modeling, the communication is not performed at a single point in time, but spread over a certain period. Depending on the requirements of the user models, either the start, the end or both start and end time of a transaction might be of interest. For this reason, the packet-level synchronization interface offers both the start-of-transfer and end-of-transfer event.

To summarize this overview of application modeling, the complete application is captured as a set of coarse grain SoC building blocks. These functional blocks are represented as reactive processes, which communicate by exchanging Abstract Data Types via a generic synchronization interface. An application model according to this representation is prepared for VAM based design space exploration.

6.2 Architecture Modeling

Ultimate goal of the MP-SoC platform design phase is the evaluation of different architectural alternatives with respect to the performance requirements. The developed VAM technology provides a set of generic architecture models, which are compliant to the generic synchronization interface. By that, VAM enables a *transparent* mapping of a given application to the anticipated architecture, i.e. the reactive processes can be mapped to any architecture configuration without further modification .

The architecture models fall into three categories, that handle different aspects of architectural elements:

- **Processing Delay Annotation** captures the fact, that the execution of a functional processes occupies a physical processing element for a certain amount of time.
- The **Virtual Processing Unit** models sharedprocessing elements, which support the multi-threaded execution of more than one functional process. This covers the impact on performance of both Software Operating Systems (SW-OS) as well as Hardware Multi-Threading (HW-MT).
- The **NoC Framework** captures the impact on performance of limited communication resources. It has been conceived to cover all kinds existing and emerging types of communication architectures like point-to-point connections, shared busses and highly complex Network-on-Chip infrastructures.

The following sections elaborate in more detail on the different architecture models.

6.2.1 Processing Delay

The annotation of processing delays is conceived to be completely orthogonal from the functional process, hence the previously validated functional system behavior is preserved. The methodology for processing delay annotation is based on the following observation: For performance profiling purposes, the basic timing characteristics of any single threaded processing element can be expressed by the temporal relationship of consuming, processing and producing tokens. This relationship is completely independent from the functionality itself, it only reflects different execution units. The performance impact of different block implementations is explored by annotating a delay time that is calculated from different timing models [25].

As depicted in figure 6.3, two cases have to be considered

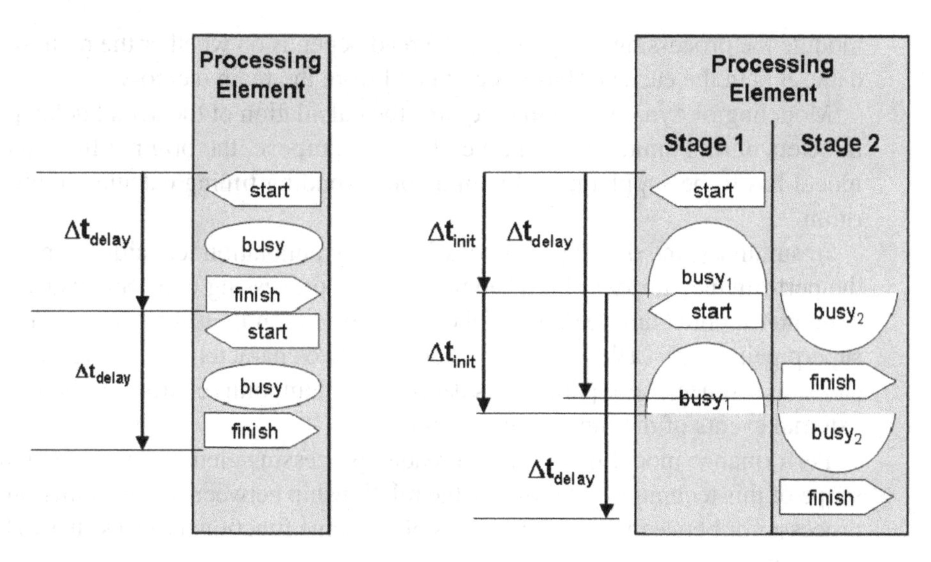

Figure 6.3. Processing Delay

- **a) Single Resource Processing Elements** are blocked for a certain amount of time during the processing of an incoming event. Thus the next event can only be consumed after the previous one is finished. An important example of single resource PEs are programmable architectures, which are busy until the current task is finished. The impact on performance of any single resource PE is completely described with a single delay parameter Δt_{delay}.

- **b) Pipelined Processing Elements** are able to consume new incoming events before the previous one is finished. This case covers the temporal properties of HW units, which employ multiple pipeline stages to improve the processing throughput. Besides the processing delay Δt_{delay}, the performance modeling of multi resource PEs requires a second timing parameter Δt_{init}, which specifies the minimum time interval between two consecutive event arrivals.

Both timing parameters can be either static or dynamic: Static delays correspond to deterministic execution times, e.g. pipelined architectures are able to consume and produce a token every cycle but introduce a static latency. This latency depends on the number of anticipated pipeline stages: the performance of a HW block with 5 pipeline stages and a clock period of $5ns$ is modeled as $\Delta t_{delay} = 25ns$ and $\Delta t_{init} = 5ns$.

In case of dynamic timing parameters the processing time depends on the actual data and the inner state of the block. For example in the case of a cache module the processing delay of a cache read depends on whether the requested data set is in the cache or has to be fetched from the main memory.

Modeling of dynamic timing requires the calculation of the actual delay parameters during simulation runtime. For this purpose, the original functional model has to be supplemented with a corresponding timing calculation algorithm.

In summary, the developed processing delay annotation technique captures the performance impact of any single threaded processing unit. Sophisticated computer architecture features of the target execution unit like superscalarity, superpipelining, VLIW schemes boil down to two characteristic timing parameters. In essence, these parameters describe the temporal relationship between external events of the functional process.

Performance modeling of multi-threaded processing elements is beyond the scope of this technique. In this case the relationship between several functional processes or between several instances of the same functional process has to be taken into account.

6.2.2 Virtual Processing Unit

In order to extend the modeling capabilities towards multi-threaded processing elements, the architecture model library comprises a generic *Virtual Processing Unit (VPU)*. The VPU has been designed to cover both Hardware Multi-Threaded (HW-MT) processing elements as well as the core functionality of Software Operating Systems (SW-OS).

The major purpose of the generic VPU is to model the impact on performance caused by the concurrent execution of more than one functional process on a single processing element. In an example depicted in the upper part of figure 6.4, two processes with their individual delay annotations are mapped to a single VPU instance. The bottom of figure 6.4 shows the resulting timing in response to an assumed scenario.

First process 1 is activated by the $start\,p_1$ event, executes the first portion of its functionality and after 10 time units generates the $request\,p_1$ event. In the meantime, the activation event $start\,p_2$ occurs, but process 2 cannot start execution before the first process is finished and swapped out. In the given scenario, the communication request from process 1 returns before process 2 has finished the first portion of its functionality. Since process 1 is configured to have a higher priority, process 2 is preempted and not resumed before process 1 has completed its functionality.

Besides modeling the timing of concurrent process execution, the generic VPU provides capabilities for automatic address resolution. This feature is indispensable for transparent process mapping to arbitrary VPU configurations. Without VPU, each communication link within the reactive process network is

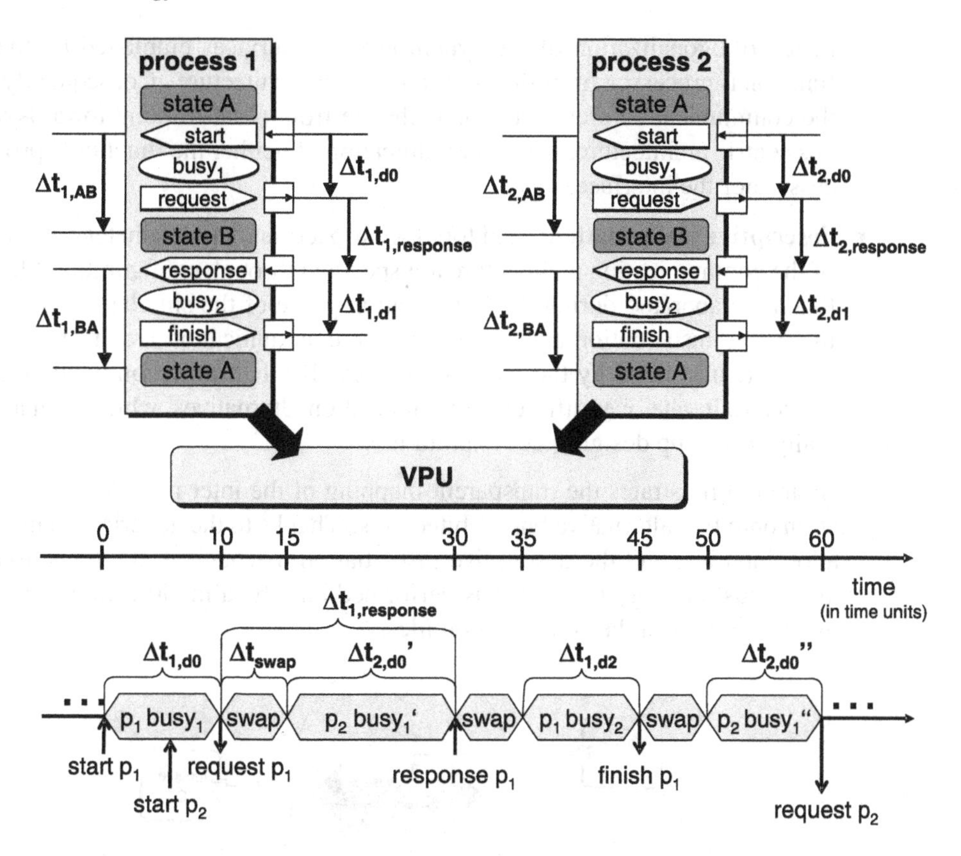

Figure 6.4. VPU Timing Calculation

unambiguously defined by the binding of the initiator output port and the target input port to a point-to-point channel. After the VPU mapping is carried out, interacting processes may be mapped to either the same or to different VPU instances. Additionally, more than one instance of a process may exist. The automatic address resolution mechanism locates the corresponding instance of the target process to preserve the original process connectivity.

6.2.3 NoC Framework

As the third aspect of MP-SoC architecture modeling, the NoC framework enables the systematic design space exploration of complex on-chip communication networks. The key idea of the NoC framework is to simplify the time-consuming process of changing communication architecture and topology. This is archived by the following two concepts:

- **generic interface:** The communication architecture is hidden behind the already introduced generic synchronization interface. This attains the complete orthogonalization of the synchronization services employed by the functional processes from the communication architecture. Consequently, the communication architecture can change from point-to-point towards a bus centric architecture, without modification of neither the functional processes nor the interface.

- **descriptive instantiation:** All topology aspects and all parameter options of the communication architecture are specified through configuration files that are elaborated during the initialization phase of the simulation. Thus the whole instantiation and binding of the communication architecture is automatically done by the NoC framework. By that, no recompilation is needed to iterate over different communication alternatives, which dramatically speeds up design space exploration.

Figure 6.5 illustrates the transparent mapping of the inter module communication onto two alternative bus architectures. Thanks to the generic synchronization interface and the descriptive instantiation mechanism, the transition from one bus topology to another is performed only by a modification of the parameter settings in the configuration files.

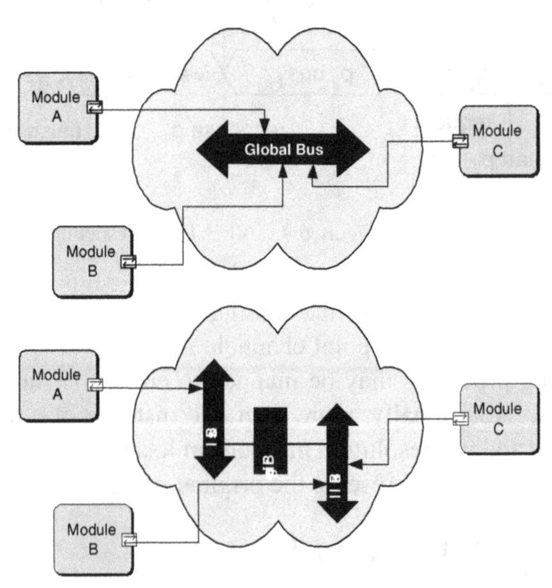

Figure 6.5. Transparent Communication Mapping

As a major differentiator against related work on communication performance modeling, the NoC framework covers not only contemporary point-to-point and shared bus architectures. Instead, the NoC framework also addresses emerging Network-on-Chip architectures, which are generally believed to provide the global on-chip communication infrastructure of future multi-processor platforms. By that, the proposed MP-SoC simulation framework and exploration methodology is designed to cope with emerging architecture trends.

6.3 Envisioned Design Flow

Now that the principles of Virtual Architecture Mapping have been introduced, this section presents the envisioned flow for the development of emerging NoC enabled MP-SoC platforms.

6.3.1 Design Flow Overview

In the first place, an overview of the complete flow is given. As depicted in figure 6.6, the overall flow is derived from the discussion on elementary design phases in section 4.2 as well as from the discussion on TLM abstraction levels in section 6.1.2.

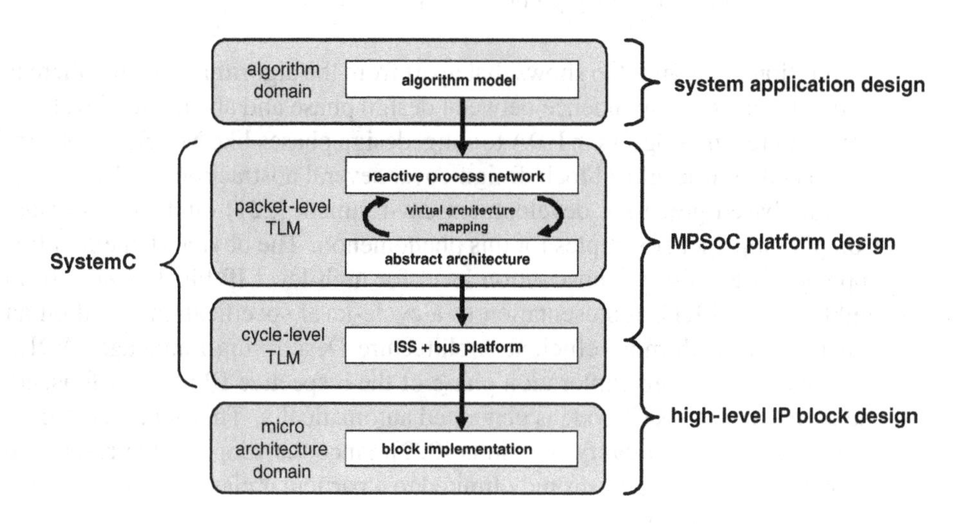

Figure 6.6. Design Flow Overview

- The *algorithm domain* directly corresponds to the functional design phase, which deals with development of Hardware independent Software and application specific algorithms.
- *Packet-level TLM* is the abstraction level for the methodology described in the course of this book. The following sections provide more details on the

individual design tasks. In essence, this level addresses the architectural exploration and application mapping of complex MP-SoC platforms.

- *Cycle-level TLM* by itself comprises several abstraction levels, which cover instruction and cycle accurate processor models as well as bus accurate and cycle callable platform models [146]. As already elaborated in section 5.2, the state-of-the-art methodology and tooling for early MP-SoC platform integration are based on this abstraction level. The major design problems addressed on this level are ISS integration, development of hardware dependent embedded software, 100% cycle accurate performance profiling, and co-simulation with implementation models for verfication purposes.

- The *micro architecture domain* covers the traditional Hardware implementation and verification flow. A Register Transfer Level (RTL) synthesizable Hardware Description Language (HDL) description of the desired functionality embodies the entry point to this domain.

Additionally, figure 6.6 shows that apart from the algorithm domain, there is no one-to-one correspondence between design phase and abstraction level. As a result of recent progress in EDA tooling, design phases like MP-SoC platform design and high-level IP block design cover several abstraction levels.

LISA based processor development environment [221] and CoWare Bus-Compiler represent examples for this phenomenon. The obvious benefit of this approach is to raise the abstraction level for high-level IP block design from highly detailed RTL representation to a cycle-level specification based on an concise and much more efficient Architecture Description Language (ADL). After the architecture exploration phase of the respective IP block is finished, the corresponding RTL code is generated automatically. This approach proves to yield convincing quality of results [222], since the scope of the respective IP creation framework is strongly limited to a particular class of IP blocks, like e.g. processors or buses.

Taking the methodology described in this book into account, the MP-SoC platform design phase covers both state-of-the-art cycle-level TLM as well as packet-level TLM. As mentioned above, the goal is to raise the abstraction level for initial architecture exploration before committing to a feasible platform configuration and spending significant effort for the creation of a cycle-level platform prototype.

Thanks to SystemC as a common specification language for both cycle-level TLM as well as packet-level TLM, the abstraction levels in MP-SoC design phase are inter-operable. This enables the designer to mix and match abstraction levels, solve individual design problems at the best suited level and successively proceed to the refined representation.

6.3.2 MP-SoC Platform Design Flow

This section now elaborates in more detail on the MP-SoC platform design phase. As depicted in figure 6.7, the discussion is focused on the design tasks performed at the packet-level TLM abstraction layer and separately treats the refinement steps of computation, functional, and communication aspects. The dashed arrows in the figure denote precedence relation between design steps, e.g. functionality must be refined to memory mapped communication before the insertion of memory models is possible.

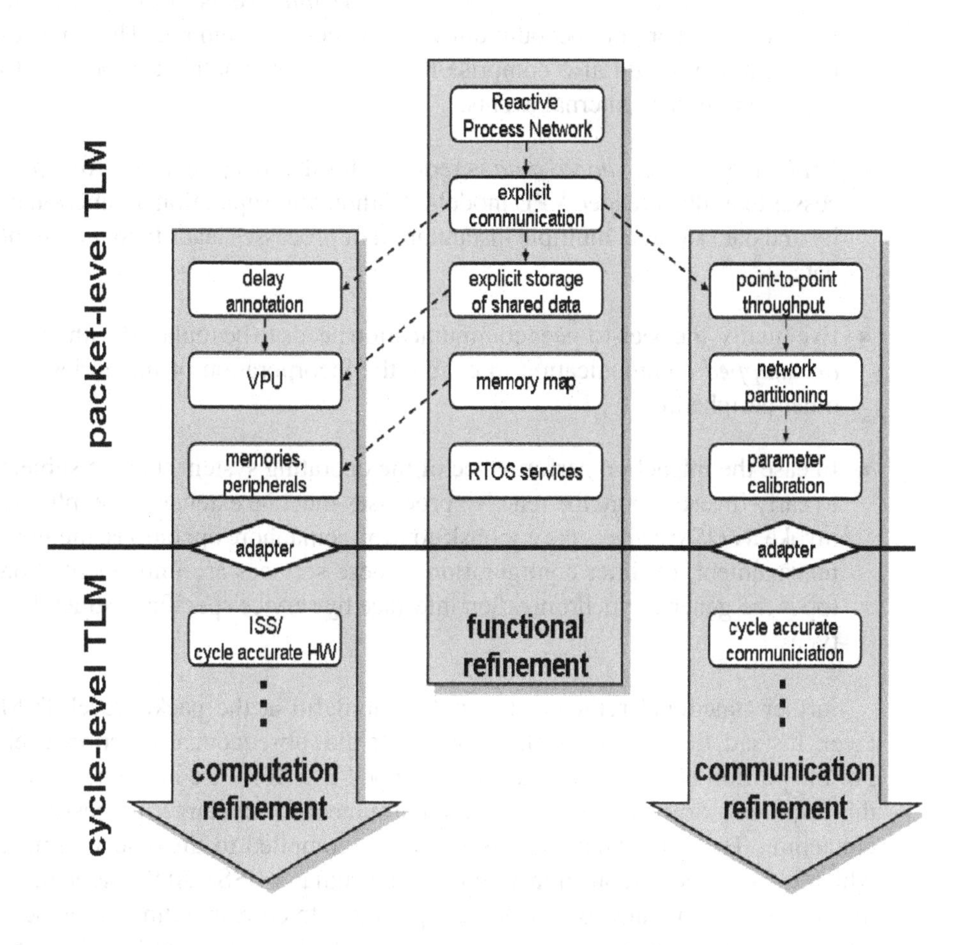

Figure 6.7. MP-SoC Design Flow

Functional Refinement. As already explained in section 6.1, the starting point of any architectural investigation is a reactive processes network of the considered application. Initially, the inside of the processes is either newly created, imported from a preceding application design phase, or can be just a non-functional workload model. During the functional refinement flow, the process network is successively developed to enable refined architecture mapping.

- *Explicit communication* is the ultimate precondition for architectural investigation. Recall, that the VAM methodology is based on the temporal sequencing of events, which occur during the simulation of the reactive processes. Consequently, only the *externally visible* events of every reactive process are taken into account during architecture mapping. This first refinement step might also comprise the splitting of functional processes to expose previously internal events.

- *Explicit storage of shared data* is required for the mapping of reactive processes to multi-threaded VPU models. Without this separation of functionality and data storage, multiple instantiation of processes leads to inconsistent data.

- Eventually, the peer-to-peer communication needs to be replaced with *memory mapped* communication to enable the incorporation of memories and other peripherals.

- In case the impact on performance of the operating system shall be subject to early investigation, the reactive processes must be extended to explicitly invoke *RTOS services*, like e.g. task control, semaphore operations, memory management, or timer configuration. These services are implemented on top if the generic synchronization interface by service specific abstract data types.

Further functional refinement is not meaningful at the packet-level TLM layer. Instead, the next step is the transition to the subsequent abstraction level. For the portion of the process network mapped to a Software implementation, this requires a Software development environment for the target processor architecture. The functional process has to be compiled to the object format, which is then executed on an Instruction Set Simulator (ISS). At the same time, the abstract communication has to be replaced with corresponding implementation by means of predefined driver libraries like in component based design [154] or by means of interface synthesis [134].

Computation Refinement. This refers to the refinement steps during the virtual architecture mapping of reactive processes to the anticipated processing elements. The individual steps are explained in sections 6.2.1 and 6.2.2 in more detail.

- The *delay annotation* models the time, which is required for the execution of the considered functionality on a physical processing element.
- *Virtual Processing Units* extend the modeling capabilities towards user defined process scheduling. By that, VPU models are essential for early consideration of multi-threaded processing elements like Hardware multi-threading or Software operating systems.
- The incorporation of *memories and peripherals* allows the early investigation of Caches, Memory Management Units (MMUs), Direct Memory Access (DMA), Interrupt Controllers (IRCs) etc. Depending on the considered application, these components have a major impact on performance and cost [218].

In case of data-plane centric applications, the timing annotation mechanism for the investigation of the anticipated computation delay proves to be very powerful in terms of accuracy and modeling efficiency [215, 218, 219, 223]. On the other hand, delay estimations in the context of highly dynamic control-plane applications is likely to yield rather vague results. In case accurate performance numbers are required, the control-plane processing portion of the application has to be profiled on the target ISS.

Communication Refinement. The NoC framework enables the systematic design space exploration of arbitrarily complex on-chip communication networks. During the simulation, the evaluation modules connected to the communication models collect statistical information like resource utilization, latency, and throughput. Based on these metrics, the system architect designs the communication infrastructure according to the following successive refinement steps:

- During the *initial throughput measurement*, the overall on-chip traffic is functionally captured by means of an unconstrained point-to-point communication. The resulting communication profile identifies interacting partners and rough throughput requirements.
- The coarse *network partitioning* is dedicated to the identification of the optimum combination of network types. The system architect maps the point-to-point communication to an appropriate set of network types by configuring the NoC framework using the generic communication models.
- By iterative *parameter calibration*, the selected communication architecture is fine-tuned to the traffic requirements. Parameters refer to e.g. the bandwidth of a bus system or the queue-length of a crossbar architecture.

Of course the network partitioning has superior impact on the final quality of results. Here the unified approach enables a rapid exploration of totally different network architectures by simply replacing the communication models. By that the system architect can optimize the communication architecture in iterative exploration cycles.

6.3.3 Mixed Level Co-Simulation

In accordance with the interface based design paradigm [70], the transition from packet-level to lower abstraction levels like cycle-level TLM or RTL is performed successively in a block by block manner. For verification purposes, the refined blocks are individually included into the abstract system context. This *Mixed Level Co-Simulation* is enabled by a generalized adapter concept, which brides the different abstraction levels with respect to data, timing and communication [215, 114, 25].

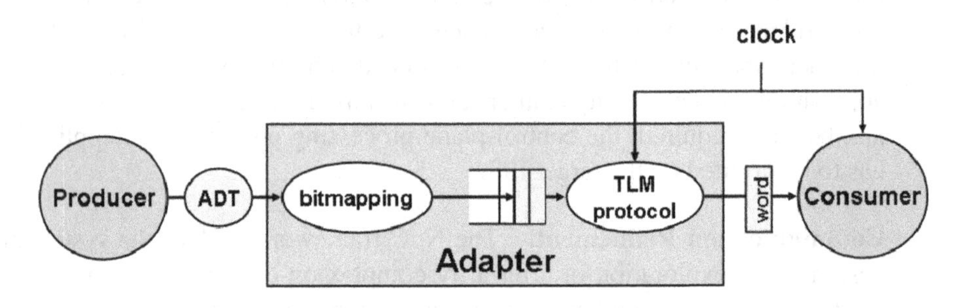

Figure 6.8. Mixed Level Co-Simulation

The concept of the mixed-level co-simulation adapter is depicted in figure 6.8, where a packet-level Producer is connected to a cycle-level Consumer module. In the first step, the bit-mapping layer in the adapter maps the ADT to the respective fields in the corresponding bit-accurate data representation. The resulting bit-stream is then transfered to the protocol layer, where it is cut into slices according to the respective data width and forwarded to the protocol layer. Here the protocol specific TLM interface methods are called to stream the data words into the consumer.

A similar adapter channel implements the reverse direction to feed the output an cycle-level producer into the packet-level consumer. Here the ADT is reconstructed from the output bit-stream provided by the protocol engine. Note that this concept is also capable to bridge packet-level and RT-level SystemC [25]. Together with commercially available SystemC-HDL Co-simulation environments, this enables the functional co-verification of VHDL or Verilog implementation against the packet-level reference model [216, 217].

Refinement of Processing Elements

At some point in the computation refinement process the achievable accuracy of Virtual Architecture Mapping is no longer sufficient to take further design and partitioning decisions. Especially in case of highly run-time dynamic control plane processing tasks the coarse grain annotation of SW execution time yields imprecise performance estimations. Despite continued research effort in this area, SW execution time analysis is still considered as an unsolved problem in HW/SW Co-design [6].

As soon as accurate information about the Software performance is required, the Software execution time has to *measured* instead of *estimated*. So far, system integration of cycle accuracy processor simulators is not performed until a cycle level TLM system model is available [146]. The mixed-level adapter allows the early integration of cycle accuracy processor simulators into the packet-level MP-SoC framework [114, 224]. By that the Software execution time can already be accurately measured during the architecture exploration phase.

6.4 MP-SoC Simulation Framework

The Virtual Architecture Mapping based system level design methodology is supported by a tool environment for simulation, debugging and analysis. Together these tools enable a systematic and fast design space exploration of complex on-chip communication networks. In this context, design space comprises both the analysis of architectural alternatives as well as the investigation of application to architecture mappings.

The overall work-flow of the MP-SoC Environment is depicted in 6.9: As a prerequisite, the considered application is represented as a functional SystemC process network. This functional model is mapped to the anticipated target architecture in order to create a performance model of the resulting system architecture. The mapping is performed virtually by instantiation and configuration of generic performance models, including the annotation of the respectively timing characteristics. By that the methodology enables a very fast exploration of totally different design alternatives. Because the process of timing annotation is completely orthogonal to the functionality, previously validated functional application models remain unaltered throughout the exploration phase.

In order to further accelerate the design space exploration cycle, the complete architecture specification is represented as a set of configuration files in eXtended Markup Language (XML) format:

- configuration of the timing model
- number of available processors and number of supported concurrent threads per processor
- mapping of the application task to processors and threads

- instantiation, parameterization and interconnection of the communication nodes
- instantiation and address mapping of the memory architecture

Before the simulation starts the architecture specification is extracted from the configuration files and bound to the application model. During the simulation, evaluation modules connected to the architecture models collect and aggregate statistical information like resource utilization, latency, and throughput. On completion of the simulation, this statistical information is visualized by means of histogram and communication graph views. Based on these data, the system architect may modify the MP-SoC architecture and/or the application mapping until the requirements are met.

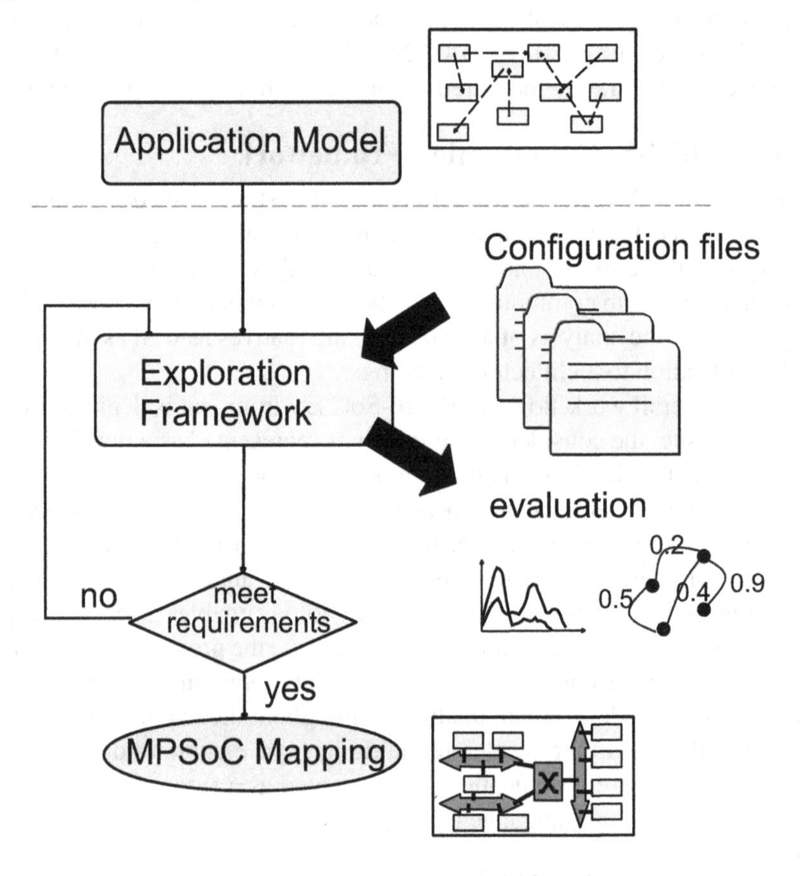

Figure 6.9. Iterative Exploration Flow

This declarative specification mechanism enables a rapid exploration of totally different communication and processing architectures by means of modification of configuration files. In case of very complex applications, the simulation driven approach can be complemented with statistical post-processing techniques like regression analysis to reduce the overall design space.

The ultimate result of the design space exploration is the *detailed specification* of the MP-SoC platform architecture as well as the application partitioning.

A more detailed description of the tools and models in the MP-SoC Simulation Framework is given in chapter 8 after the definition of the underlying timing model.

Chapter 7

UNIFIED TIMING MODEL

This chapter provides a formal definition of the timing model, which constitutes the underlying technology of the Virtual Architecture Mapping methodology. This timing model enables the joint consideration of functionality and architecture by successively annotating the timing characteristics of the anticipated architecture to the application model. In essence, the timing model describes the mapping of the un-timed *functional* events in the application model to the timed *physical* events in the anticipated architecture.

The description of the denotational semantics of the timing model relies on the notation proposed by the Tagged Signal Model (TSM) framework [26], which provides a widely accepted formalism to reason about the characteristics of Models of Computation [225]. The basic denotational TSM specification is supplemented with the operational semantics of the timing model to create the link to the corresponding simulation environment, which is described in the following chapter.

The first part of this chapter briefly recalls the major constructs of the Tagged Signal Model framework, which are then related to the context of event-driven SystemC simulation semantics at the TLM level. Afterwords, the individual aspects of the timing model are introduced, covering step by step the initial functional application model, as well as the delay annotation caused by processing elements, virtual processing units, and communication resources. This chapter closes with the specification of the performance metrics, which can be derived from the timing model.

7.1 Tagged Signal Model Introduction

This section first recalls basic definitions from the Tagged Signal Model framework [26] to introduce the employed notation. Additionally, major properties of the well-known event-driven simulation Model of Computation are replicated to serve as a foundation for the subsequent discussion of the timing model.

7.1.1 Elementary Definitions

The original motivation behind the Tagged Signal Model (TSM) is to serve as a *denotational framework (a 'meta model') within which certain properties of models of computation can be compared* [26]. The TSM precisely defines general terms like e.g. process, signal, and event, which are present in any MoC. Based on these terms, general MoC properties like determinism, causality and synchronization can be derived and compared.

The fundamental entity in the TSM is an event, i.e. a value/tag pair. Tags are often used to denote temporal behavior and sets of events are aggregated into a signal. Processes are relations on signals, expressed as sets of n-tuples of signals. A particular model of computation is distinguished by the order it imposes on tags and the character of processes in the model.

Given a set of *values* $\mathcal{V}$ and a set of *tags* $\mathcal{T}$, an *event* e is a member of $\mathcal{T} \times \mathcal{V}$, i.e. an event has a tag and a value. A *signal* s is a set of events, which can be viewed as a subset of $\mathcal{T} \times \mathcal{V}$, or as a member of the power-set $s \in \wp(\mathcal{T} \times \mathcal{V})$. A functional signal is a (possibly partial) function from $\mathcal{T}$ to $\mathcal{V}$.

The set of all signals is denoted $\mathcal{S}$. A tuple of n signals is denoted s, and the set of all such tuples is denoted $\mathcal{S}^n$. The position of a signal in a tuple is denoted by an index and uniquely identifies an individual signal.

In this context, a *process* $\mathcal{P}$ is described as a set of N signals, i.e. $\mathcal{P} \subseteq \mathcal{S}^n$. A particular $s \in \mathcal{S}^n$ corresponds to the *behavior* of a process if $s \in \mathcal{P}$. The *composition* of processes is defined as the cross product of the set of behaviors. In case there is no interaction between the processes, the composition is simply a superposition of both behaviors as depicted in the left part of figure 7.1. On the other hand, one or more signals of different processes may be identical. In this case, each pair of identical signals constitutes a *connection* C between the processes. The connection $C \subset \mathcal{S}^n$ itself is defined to be a process, where two or more signals of the n-tuple are constrained to be identical. Thus, $C_{j,k}$ defines a connection in $s \in \mathcal{S}^n$, such that $s_j = s_k$ holds.

The signals of a process can be further distinguished into a set of N_I input signals $\mathcal{I}$ and a set of N_O output signals $\mathcal{O}$. This separation into inputs and outputs can be described with a mapping π. Let $I = (i_1, \ldots, i_m)$ be an ordered set of indices in the range $1 \leq i \leq n$, and define a *projection* $\pi(s)$ of $s = (s_1, \ldots, s_n) \subseteq \mathcal{S}^n$ onto $\mathcal{S}^m$ with $m \leq n$ by $\pi_I(s) = (s_{i_1}, \ldots, s_{i_m})$. By that, the ordered set of indexes I defines the signals that are part of the projection and the order in which they appear in the resulting tuple. In order to describe the generation of events on the outputs of a process depending on its inputs, an ordered set of indices I for the N_I input signals and an ordered set of indices O for the N_O input signals is given.

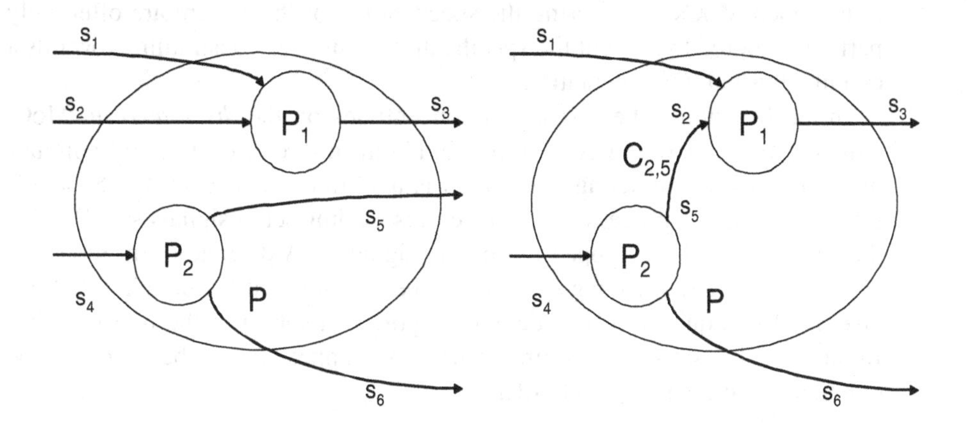

Figure 7.1. Composition of Processes

In the example depicted on the left side of figure 7.1, the input index set is $I = (1, 2, 4)$ and the output index set is $O = (3, 5, 6)$. By that, the input signals are constrained to be $\mathcal{I} = \pi_I(s) = (s_1, s_2, s_4)$ and the output signals are $\mathcal{O} = \pi_O(s) = (s_3, s_5, s_6)$.

A process is *determinate* if for any input it has exactly one behavior or exactly no behaviors, otherwise it is *nondeterminate*. A process in $\mathcal{S}^n$ that is functional with respect to (I, O) is obviously determinate if I and O together contain all the indexes in $1 \le i \le n$.

A process is *functional* with respect to (I, O) if for every $s \in \mathcal{P}$ and $s' \in \mathcal{P}$ where $\pi_I(s) = \pi_I(s')$, it follows that $\pi_O(s) = \pi_O(s')$. For such a process, there is a single-valued mapping $f : \mathcal{S}^{N_I} \mapsto \mathcal{S}^{N_O}$, such that for all $s \in \mathcal{P}$, $\pi_O(s) = f(\pi_I(s))$. A process is *total* if f is defined over all $\mathcal{S}^{N_O}$, otherwise it is *partial*.

If all processes are functional with inputs on the left and outputs on the right, then obviously the composition processes are also functional and thus preserves determinacy. A much more complicated situation involves feedback. Whether determinacy is preserved in the presence of feedback loops depends on the tag system and more details about the process.

7.1.2 Tag Systems

The tag system employed in the Model of Computation together with the ordering relation among the tags is responsible for synchronization, causality, and determinism. In timed MoCs, the tags are totally ordered and can be naturally interpreted to mark the time in a physical system. By that, signals can be interpreted as a temporal sequence of events. On the other hand, tags in un-timed MoCs used during the specification of the system are often only partially ordered to prevent the specification from over-constraining towards a certain physical implementation.

In the following, the discussion concentrates on the *discrete-event* MoC, which is the underlying simulation paradigm in circuit design and communication network modeling and which is also implemented by the SystemC simulator. For any process $\mathcal{P}$ and the corresponding set of signals $s \in \mathcal{P}$, $T(s)$ denotes the set of tags appearing in any signal s. A discrete-event model of computation has a timed tag system, i.e. the tags are totally ordered. Additionally, $T(s)$ is required to be order isomorphic to a subset of the integers [26]. Intuitively, this says that the time stamps that appear in any behavior can be enumerated in chronological order.

Causality is a key concept in the design of discrete-event simulators. The problems center around how to deal with synchronous events (those with identical tags) and how to deal with feedback loops. Lee defines a suitable metric to describe the temporal relation of discrete signals:

$$d(s, s') = sup\{\frac{1}{s^t} : s(t) \neq s'(t), t \in T(s) \cap T(s')\} := \frac{1}{s^\tau}$$

According to this definition, τ corresponds to the the smallest tag where the two signals s and s' differ. In case s and s' are identical, τ is infinite and $d(s, s') = 0$.

A sequence of events on a signal is called Cauchy sequence in case the sequence of events converges towards a limit within the metric space, i.e. there exists for any $\epsilon > 0$ a natural number $n_0(\epsilon)$ such that $d(s(t_n), s(t_m)) < \epsilon$ for all $n, m \geq n_0(\epsilon)$. A metric space is complete if every Cauchy sequence of points in the metric space that converges does so to a limit that is also in the metric space.

Based on this metric, Lee classifies three different forms of causality.

DEFINITION 7.1 (CAUSALITY) *A function $f : \mathcal{S}^m \mapsto \mathcal{S}^n$ is* causal *if for all input signals $s, s' \in \mathcal{S}^m$*

$$d(f(s), f(s')) \leq d(s, s')$$

Intuitively, two possible outputs differ no earlier than the inputs that produced them.

DEFINITION 7.2 (STRICT CAUSALITY) *A function $f : \mathcal{S}^m \mapsto \mathcal{S}^n$ is* strictly causal *if for all input signals $s, s' \in \mathcal{S}^m$*

$$d(f(s), f(s')) < d(s, s')$$

DEFINITION 7.3 (DELTA CAUSALITY) *A function* $f : \mathcal{S}^m \mapsto \mathcal{S}^n$ *is* delta causal *if there exists a positive real number* $0 \leq \delta < 1$ *such that for all input signals* $s, s' \in \mathcal{S}^m$

$$d(f(s), f(s')) < \delta d(s, s')$$

Intuitively, this means that there is a strictly positive number, before any output of a process can be produced in reaction to an input event. The latter inequality is recognizable as the condition satisfied by a *contraction mapping*.

For a complete metric space, the *Banach fixed-point theorem* [226] states, that if $f : \mathcal{S}^n \mapsto \mathcal{S}^n$ is a contraction mapping, then there is exactly one $s \in \mathcal{S}^n$ such that $s = f(s)$.

The Banach fixed-point theorem provides the means to reason about causality and determinism during the composition of processes in the presence of feedback loops: if the process is functional and delta causal, then the feedback loop has exactly one behavior (i.e. it is determinate). Additionally, it defines a constructive way to find the fixed point by means of an iterative evaluation algorithm, which is indeed implemented in discrete-event simulators.

7.1.3 SystemC Simulation Tag System

Similar to the VHDL simulation paradigm [227], the SystemC simulation kernel [140] implements the following iterative simulation cycle, which corresponds to the iterative nature of the Banach fixed-point theorem. The reader may refer to Mueller et al. [228] for the operational semantics of the SystemC language elements.

1 *Evaluate Phase* From the set of processes that are ready to run, select a process and resume its execution. The order in which processes are selected for execution from the set of processes that are ready to run is unspecified. The execution of a process may generate new events, which are pending until the update phase. Pending events are reported to the SystemC kernel by calling the `request_update()` function.

2 Repeat Step 1 for any other processes that are ready to run.

3 *Update Phase* All pending events generated during the previous evaluate phase are updated and eventually become active events. Event activation is reported to the SystemC kernel by calling the `notify()` function.

4 If there are newly activated events, determine which processes are ready to run and go to step 1.

5 If there are no more events, the simulation is finished.

6 Else, advance the current simulation time to the time of the earliest pending timed event.

7 Determine which processes become ready to run due to the events that have pending notifications at the current time. Go to step 1.

The simulation time does not advance before step 6, hence the alternating evaluate and update phases are executed at a fixed point in global simulation time. By that steps 1 – 4 represent the iterative contraction mapping, which determines the final status of the processes at any point in time. Conceptually, this evaluate-update synchronization mechanism implements a two dimensional tag system $\mathcal{T} = \mathbf{N}^+ \times \mathbf{N}^+$, where $\mathbf{N}^+$ denotes the set of all positive natural numbers. By that, each event $e \in \mathcal{T} \times \mathcal{V}$ carries a time tag $t = (t_1, t_2)$, which is composed of an *absolute simulation time* t_1 and a *delta delay* t_2. The ordering relation between tags $t \preceq t^{'}$ is defined if $t_1 < t_1^{'}$ or if $t_1 = t_1^{'}$ and $t_2 < t_2^{'}$. Note, that the delta time is not exposed to the user, but is merely maintained in the SystemC simulation kernel.

Unfortunately, the $\mathbf{N}^+ \times \mathbf{N}^+$ does not guarantee delta-causality, since there can be an infinite number of tags between two tags. Therefore the user is responsible to prevent from creating feedback-loops, which would result in infinite execution of simulation steps 1–4. At least the delta delays in the SystemC simulator are sufficient to ensure determinacy, but not enough to ensure that a feedback system has a behavior at all.

7.1.4 Cycle-Level Tag Systems

In the context of cycle-level Transaction Level Modeling, there is an opportunity to employ different synchronization schemes on top of the discrete-event SystemC simulation kernel to ensure deterministic behavior. As one popular example the two-phase synchronization scheme has been conceived for efficient modeling and simulation of bus centric SoCs [18]. It is based on the assumption, that all master processes and the bus are synchronized to a single clock. The master processes generate communication requests to the bus during the activation of the *positive* edge of the clock. These requests are gathered by the bus, which is then activated at the *negative* edge of the clock to arbitrate the requests.

Effectively, this two-phase synchronization scheme reduces the MoC to a one dimensional tag system $\mathcal{T} = \mathbf{N}^+$. On the one hand side, this restricts the usable language constructs and requires a clocked register within any feedback loop in the system. On the other hand, this enables to replace the general event-driven simulation kernel with a much more simulation speed efficient methodology specific cycle-driven kernel.

As an important restriction, this two-phase synchronization scheme can only cope with a single arbitration node between the masters and the slaves. In case of more complex bus-matrix architectures, one negative clock edge is not enough to process multiple bus nodes like input-stages, output-stages and the central bus. In this case, a combinatorial processing of all bus requests is required to determine the final state of the bus.

7.1.5 Network Simulators

A commercial discrete-event simulator in the networking application domain is OPNET [220], which serves as an design environment for telecommunication networks. OPNET is based on a real-valued tag system and provides three hierarchical modeling levels: the macroscopic network level, the local node level and the internal process level. Synchronous events are only supported at the network and node level, whereas events at the process level are inserted into a single FIFO event queue. Events with identical tags result in subsequent activations of the target process. By that, the state transitions in the process depend on the non-deterministic sequence in the FIFO queue.

The implicit non-determinism significantly hinders modeling of synchronous events at the process level, which is essential for the modeling of synchronous circuits. For this reason, traditional network simulators are not suitable for the evaluation of Network-on-Chip architectures, which are tightly integrated into a synchronous hardware context. Therefore, the NoC modeling environment described in this book is implemented on top of the SystemC simulation kernel.

TSM Summary

The Tagged Signal Model developed by E.A. Lee and A. Sangiovanni-Vincentelli provides a very general framework for reasoning about the properties of Models of Computation. So far, this chapter has recalled basic definitions and major results to understand the event-driven SystemC simulation kernel with respect to synchronization, causality and determinism.

In the following, the discussion turns to the timing model underlying the system-level design of heterogeneous HW/SW multiprocessing systems.

7.2 Reactive Process Network

This section formally defines the construction of the application model[1] in terms of a Reactive Processes Network, which constitutes the starting point for subsequent architecture mapping. The notation as well as the properties of an

[1] Please refer to section 6.1 on page 60 for an intuitive introduction on application modeling

event-driven simulation are adopted from the Tagged Signal Model formalism. Additionally, the operational semantics for the manipulation of event tags are specified in a C-like pseudo syntax.

After a brief definition of common data structures, this section deals with the definition of a Functional Process, which models the behavior of the user application. The latter part introduces the Reactive Channels and describes the construction of the Reactive Process Network as a set of Functional Processes, which communicate via a set of Reactive Channels.

Common Data Structures

The Reactive Channel as well as the later defined timing manipulation nodes employ two elementary data structures, which are defined in the following together with the available access methods. First the *FIFO Queue* stores data in First-In-First-Out order.

DEFINITION 7.4 (FIRST-IN-FIRST-OUT QUEUE) *A FIFO Queue* $\mathcal{U}_{FIFO}$ *is a set arbitrary data items* $u_i \in \mathcal{U}$ *and the following access methods:*

- *enqueue() appends a new data item at the end of the list.*
- *dequeue() returns the head-of-list data item and removes it from the list.*
- *length() returns the number of data items in the list.*

Apart from the well-known FIFO Queue, the *Delay Queue* is a specific data structure for the purpose of projecting events into the future. Events in the Delay Queue are sorted according to the chronological order of the event tags. Each tag corresponds to the deadline at which the dequeue operation removes the respective event from the queue. This concept is inspired by the priority queue solution for the Pending Event Set (PES) problem [229, 230] in discrete event simulations [231].

DEFINITION 7.5 (DELAY QUEUE) *A Delay Queue* $\mathcal{U}_{DQ}$ *is a set of events* $e_i \in \mathcal{T} \times \mathcal{V}$*, which are sorted according to their tags. The event tags are increasing from head to tail, i.e. later deadlines are inserted at the end of the queue. A delay queue provides the following access methods:*

- *insert() sorts a new event into the Delay Queue according to the chronological order of the event tags.*
- *readFirst() returns the head-of-list event without removing it from the list.*
- *dequeue() returns the head-of-list event and removes it from the list.*
- *isEmpty() returns true in case the Delay Queue is empty. Otherwise it returns false.*

These two data structures are employed in the subsequent definitions of the timing model.

Functional Process

The Functional Process models the sequential behavior of a certain component and by that represents the atomic entity in a parallel process network. The

Functional Process is constructed from a Communicating Extended Finite State Machine (CEFSM) modeling a sequential portion of the user application, and the external interface according to the elementary definitions in section 7.1.1.

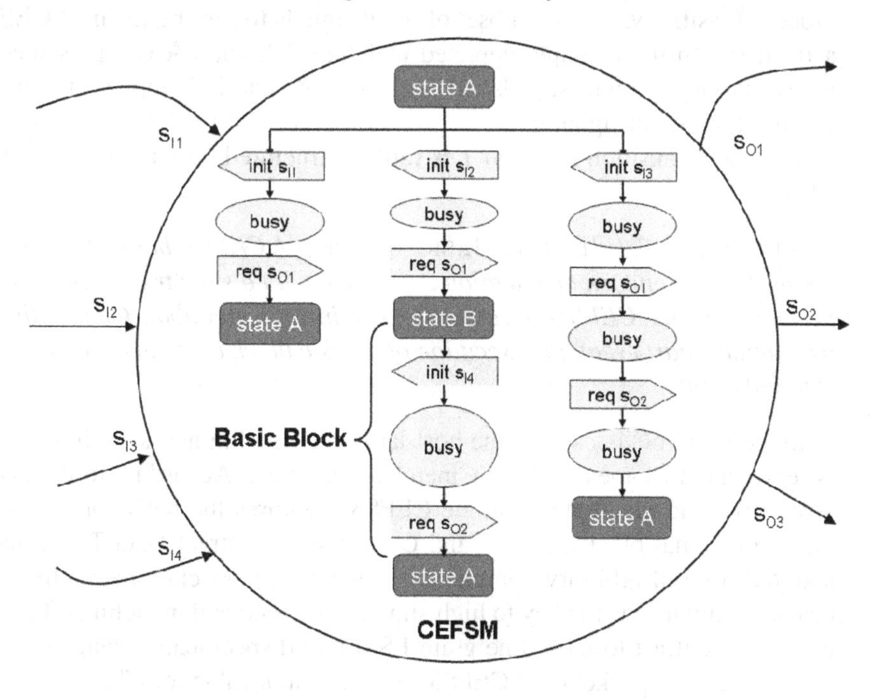

Figure 7.2. Functional Process with CEFSM

As depicted in the center of figure 7.2, the user view of a Functional Process is described as a reactive Communicating Extended Finite State Machine (CEFSM). The CEFSM is activated on the arrival of a new input symbol and responds with a state transition and eventually the generation of one or more output symbols.

DEFINITION 7.6 (COMMUNICATING EXTENDED FINITE STATE MACHINE) *A CEFSM is a 6-tuple* $(\mathcal{Z}, z_0, \mathcal{I}, f, \mathcal{O}, \mathcal{U})$ *with*

- *a finite, non-empty set of explicit states* $\mathcal{Z}$
- *a starting state* $z_0 \in \mathcal{Z}$
- *a set of input symbols* $\mathcal{I}$
- *a set of output symbols* $\mathcal{O}$
- *a state transition function* f*, where*

$$f : \mathcal{Z}^* \times \mathcal{I} \mapsto \mathcal{Z}^* \times \mathcal{O}$$

- *a set of variables* $\mathcal{U} = (u_1, \ldots)$*, which represent the implicit state*

$\mathcal{Z}^*$ denotes the global state $\mathcal{Z}^* = \mathcal{Z} \times \mathcal{W}(u_1) \times \ldots$, where $\mathcal{W}(u_i)$ denotes the possible values of variable $u_i \in \mathcal{U}$.

In the context of this book, the explicit state $\mathcal{Z}$ is represented by the actual process sensitivity, i.e. the subset of input signals triggering the next CEFSM activation. In the example depicted in figure 7.2, state A is represented by the sensitivity to input signals s_{I1}, s_{I2}, s_{I3} and state B is represented by the sensitivity to input signal s_{I4}.

The state transition function f is further structured into a number of basic blocks:

DEFINITION 7.7 (CEFSM BASIC BLOCK) *A CEFSM basic block denotes the processing and communication associated with a state transition, which is performed by the CEFSM in response to an incoming symbol. Once activated by a unique start event, the execution of a basic block is not dependent on any external event.*

Since SystemC is used as the host language, the implicit state information is represented by means of C++ member variables. According to the packet level modeling paradigm, a single CEFSM captures the behavior of coarse-grain functional blocks, thus $\mathcal{I}$ and $\mathcal{O}$ represent Abstract Data Types and $\mathcal{U}$ and f can be of arbitrary complexity, like e.g. a C++ class hierarchy. This coarse granularity is the key to high simulation speed and modeling efficiency, but — in contrast to more fine grain FSM based specification languages like e.g. Esterel — a packet level CEFSM is hardly applicable for RTL synthesis or formal analysis.

Now a Functional Process is defined as a composition of the external interface and a CEFSM representing the behavior.

DEFINITION 7.8 (FUNCTIONAL PROCESS) *Given a set of $n = N_I + N_O$ signals $\mathcal{S}^n$. A Functional Process $\mathcal{P}_{FP} \in \mathcal{S}^n$ is a process given by the 3-tuple $\mathcal{P}_{FP} = (\mathcal{S}_{FP,I}, \mathcal{S}\mathcal{FP}, \mathcal{O}, CEFSM_{FP})$ with*

- *a set of N_I input signals $\mathcal{S}_{FP,I} = (s_{i_1}, \ldots, s_{i_{N_I}})$*
- *a set of N_O output signals $\mathcal{S}_{FP,O} = (s_{o_1}, \ldots, s_{o_{N_O}})$*
- *a Communicating Extended Finite State Machine $CEFSM_{FP}$ according to definition 7.6 modeling the behavior of the Functional Process.*

In the context of packet level TLM system modeling, the values carried by the input and output events of any process are constrained to be Abstract Data Types, i.e. $v \in \mathcal{V}_{ADT}$.

So far, a Functional Process produces the events on the output signals at exactly the same point in time as the corresponding input event. Thus, a Functional Process is not strictly causal according definition 7.2.

Reactive Channel

The Reactive Channel is the crucial element to achieve strict causality and determinism for the construction of networks of functional processes. For this purpose, the Reactive Channel exactly follows the discrete-event simulation MoC explained in section 7.1.3. Hence, the properties with respect to causality and determinism of the resulting Reactive Processes Network are inherited from the two dimensional tag system $\mathcal{T} = \mathbf{N}^+ \times \mathbf{N}^+$.

The implementation of the Reactive Channel is similar to the `sc_signal` channels of the SystemC library, which are used for RTL modeling: The Reactive Channel provides variables to store projected and current values. This separation of the current and the projected value is required for the delta-cycle based event synchronization [2]:

During an activation of the Functional Process in the evaluate phase of the simulation cycle, the Functional Process produces new output events. These events are stored in the projected value of the Reactive Channel. During the subsequent update phase, the projected value is propagated to the current value. In the next evaluation phase the consumer process is activated and can read this current value.

For the sake of modeling efficiency at higher abstraction levels, the Reactive Channel differs in two aspects from the VHDL like `sc_signal` channel. First, a Reactive Channel also activates the consumer process in case the same Abstract Data Type is written again. Instead, writing the same bit to a hardware signal does not change the logic level and hence does not cause a process activation.

Second, a producing Functional Process may generate more than one output event during a single activation. In this case, the Reactive Channel temporarily stores the values in an internal FIFO queue. Later the values are forwarded to the output signal during the subsequent delta cycles of the SystemC simulation. Of course hardware signals cannot store any values, so in case of the `sc_signal` channel multiple events generated during a single process activation overwrite each other and only the last value is forwarded to the consumer process.

DEFINITION 7.9 (REACTIVE CHANNEL) *Given a set of 2 signals $\mathcal{S}^2$. A Reactive Channel $\mathcal{P}_{RC} \in \mathcal{S}^n$ is a process given by the 5-tuple $\mathcal{P}_{RC} = (s_{RC,I}, s_{RC,O}, \mathcal{U}_{RC,projected}, u_{RC,current}, f_R$ with*

- *one input signal $s_{RC,I}$,*
- *one output signal $s_{RC,O}$,*
- *a set of state variables $\mathcal{U}_{RC,projected} \subseteq \mathcal{V}_{ADT}$, which store projected values in First-In-First-Out order according to definition 7.4,*
- *a state variable $u_{RC,current} \in \mathcal{V}_{ADT}$, which stores the current value,*

[2]Please refer also to section 7.1.3 on page 83

- *a state transition function $f_{RC,activate}$, which is executed on the arrival of incoming events and appends the value v_I of the respective input event e_I to the projected value list:*

  $f_{RC,activate}\{$
  $\mathcal{U}_{RC,projected}$*.enqueue($v_I$);*
  request_update();
  $\}$

- *a state transition function $f_{RC,update}$, which is executed during the update phase of the SystemC simulation kernel. It assigns the projected value to the current value and creates an event e_O on the output signal:*

  $f_{RC,update}\{$
  $u_{current} = \mathcal{U}_{projected}$*.dequeue();*
  e_O*.notify();*
  if ($\mathcal{U}_{projected}$.length() > 0) { request_update(); }
  $\}$

As long as the input signal is functional, each event on the input signal $e_I = (t, v_I) = ((t_{I1}, t_{I2}), v_I)$ produces one event on the output signal $e_O = (t_O, v_O) = ((t_{I1}, t_{I2} + 1), v_I)$, i.e. events are delayed by one delta cycle of the SystemC simulation kernel. Non-functional input events are delayed by more than one delta cycle. By that, the composition of a Reactive Process and a set of Reactive Channels connected to each of the outports establishes a functional and strictly causal Reactive Process.

A network of Functional Processes connected via Reactive Channels is not yet sufficient to create a functional model of any embedded functional model. The concept of an intrinsic activation condition is necessary to model stimuli generators as well as self activating functional blocks, like e.g. an arbitration unit.

In this context, the notion of intrinsic activation can be represented by a self-loop Reactive Delay Channel, where Δt_{ij} captures the time between self-activating events e_i and e_j.

DEFINITION 7.10 (REACTIVE DELAY CHANNEL) *Given a set of 2 functional signals $\mathcal{S}^2$. A Delay Activation Channel $\mathcal{P}_{RDC} \in \mathcal{S}^2$ is a functional process given by a 4-tuple $\mathcal{P}_{RDC} = (s_I, s_O, \mathcal{D}_{RDC}, f_{RDC,activate})$*

- *one input signal s_I,*
- *one output signal s_O,*
- *a set of delays $\mathcal{D}_{RDC} = \{\Delta t_{delay,i}\}, \Delta t_{delay,i} \in \mathbf{N}^+$*
- *a state transition function $f_{RDC,activate}$, which is executed on the arrival of incoming events and notifies the output event with respect to the specified delay.*

  $f_{RDC,activate}\{$
  e_O*.notify(Δt_{delay});*
  $\}$

Effectively, for each event on the input signal $e_I = (t_I, v_I) = ((t_{I1}, t_{I2}), v_I)$ the Reactive Delay Channel $\mathcal{P}_{RDC}$ produces one event on output signal $e_O = (t_O, v_O) = ((t_{I1} + \Delta t_i, 0), v_I)$, i.e. events e_i are delayed by the period Δt_i.

Now all required components are defined to construct an application model in terms of a Reactive Process Network.

As depicted in the example in figure 7.3, the Reactive Process Network is constructed by connecting the input signals of the Reactive Channels to the output signals of the Functional Processes and vice versa. The notion of absolute time is only introduced by the self-loop Reactive Delay Connections, which realize intrinsic delayed activation. This kind of delayed self-activations are required to model e.g. workload generators in a test-bench, or TDMA arbitration modules.

DEFINITION 7.11 (REACTIVE PROCESS NETWORK) *A Reactive Process Network is a 6-tuple* $\mathcal{N}_{RPN} = (\bar{\mathcal{P}}_{FP}, \bar{\mathcal{P}}_{RC}, \bar{\mathcal{P}}_{RDC}, \mathcal{M}_I, \mathcal{M}_O, \mathcal{M}_A)$, *where*

- $\bar{\mathcal{P}}_{FP}$ *is a set of Functional Processes* $\mathcal{P}_{FP}$ *according to definition 7.8*
- $\bar{\mathcal{P}}_{RC}$ *is a set of Reactive Channels* $\mathcal{P}_{RC}$ *according to definition 7.9*
- $\bar{\mathcal{P}}_{RDC}$ *is a set of Reactive Delay Channels* $\mathcal{P}_{RDC}$ *according to definition 7.10*
- $\mathcal{M}_I$ *is a bijective mapping, that assigns each input signal* $s \in \mathcal{S}_{FP,I}$ *of each Functional Process* $\mathcal{P}_{FP} \in \bar{\mathcal{P}}_{FP}$ *to the output signal* $s_{RC,O}$ *of one Reactive Channel* $\mathcal{P}_{RC} \in \bar{\mathcal{P}}_{RC}$,
- $\mathcal{M}_O$ *is a bijective mapping, that assigns each output signal* $s \in \mathcal{S}_{FP,O}$ *of each Functional Process* $\mathcal{P}_{FP} \in \bar{\mathcal{P}}_{FP}$ *to the input signal* $s_{RC,I}$ *of one Reactive Connection* $\mathcal{P}_{RC} \in \bar{\mathcal{P}}_{RC}$
- $\mathcal{M}_D$ *is a bijective mapping, that assigns the input/output signal pair of a Functional Process* $\mathcal{P}_{FP} \in \bar{\mathcal{P}}_{FP}$ *to one output/input signal pair of a Reactive Delay Connection* $\mathcal{P}_{RDC} \in \bar{\mathcal{P}}_{RDC}$

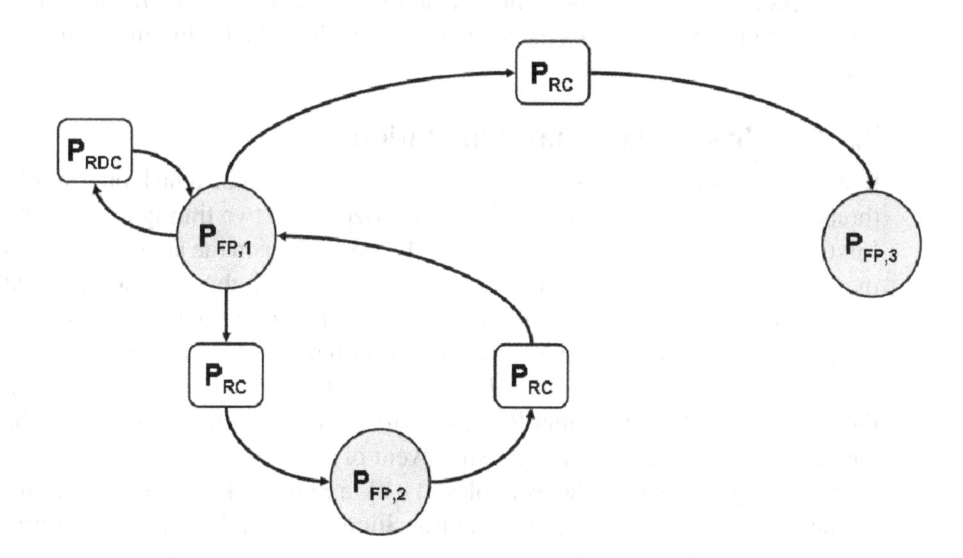

Figure 7.3. Reactive Process Network

The causal and deterministic behavior of the Reactive Processes Network in the presence of zero-delay feedback loops is ensured by the Reactive Channels. These are implemented on top of the like request/update synchronization scheme, which is provided by the SystemC simulator. Hence, the simulation semantics are similar to any HDL simulator and therefore intuitive for any user familiar with hardware modeling experience. Still the user is responsible to prevent from creating infinite feedback loops in the application model[3].

The major purpose of the Reactive Process Network is to specify the processing and communication requirements of the application. After this initial specification, the Reactive Process Network generates the *functional* events of the application, which have to be mapped to the *physical* events of the anticipated architecture during the subsequent Virtual Architecture Mapping phase.

7.3 Architecture Model

Starting from the application model, this section describes the creation of an abstract architecture model. In essence, this mapping is performed by inserting *timing manipulation nodes* into the initial Reactive Process Network. These timing manipulation nodes capture the impact on performance of the anticipated target architecture executing the application. The timing manipulation nodes are closely related to the aspects of Virtual Architecture Mapping, which are introduced in section 6.2: specialized nodes handle processing delay annotation, virtual processing units, and communication architectures. The remainder of this section formally defines the semantics of each of the timing manipulation nodes.

7.3.1 Processing Delay Annotation

Recalling the discussion in section 6.2.1, the execution of a task on a single threaded processing element can be characterized with two timing parameters: First, the processing delay Δt_d describes the time between the arrival event and the event(s) produced in reaction to this arrival. Second, the initiation interval Δt_i describes the minimum time between two consecutive arrival events.

As illustrated by the examples in figure 7.4, initiation intervals and processing delays are annotated to the *externally visible* events of the CEFSM representing the behavior of the Functional Process. More precisely, the timing has to be annotated with respect to the activating event of a CEFSM basic block[4].

According to this rule, the examples a) – c) in figure 7.4 represent legal timing annotations. In contrast, the state transition of example d) is not conform

[3]Please refer to section 7.1.3 on page 83
[4]according to definition 7.7 on page 88

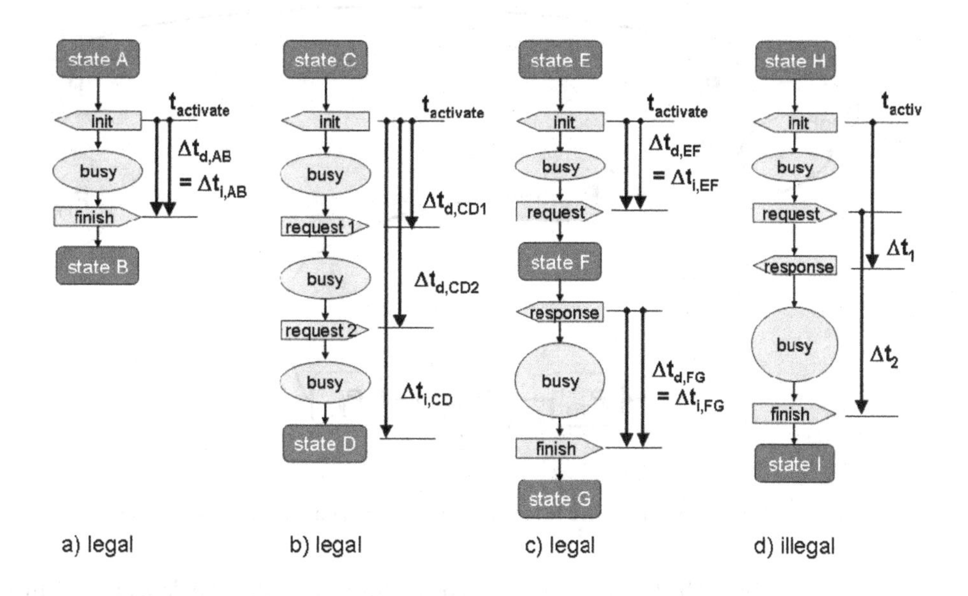

Figure 7.4. CEFSM Timing Annotation Examples

with the definition of a CEFSM building block, since the response represents an intermediate input event, which requires the insertion of an additional state between the request and the response event. Hence also the timing annotations are illegal: Both Δt_1 and Δt_2 try to define a temporal relation between events from different basic blocks. Intuitively, Δt_1 and Δt_2 are no proper timing annotations, because the time between the request and response events depends on the system context. Example c) resolves this the erroneous annotation by inserting an explicit state.

As depicted in the example in figure 7.5, an existing Reactive Process Network is augmented with timing annotations by inserting Initiator Nodes at the output signals and Target Nodes at the input signals of the Functional Processes respectively. The delay caused by the Initiator Nodes already ensures causal and deterministic behavior, so the Reactive Channels are no longer necessary.

Initiator Node. An *Initiator Node* is connected to each of the output signals of a Functional Process and implements the annotation of a processing delay. Similar to the Reactive Channel, the Initiator Node internally maintains a list of all projected events, where all of the output events generated by the Functional Process are temporarily stored. As a major difference, the events are not stored in a FIFO Queue, but sorted into a Delay Queue in chronological order of the

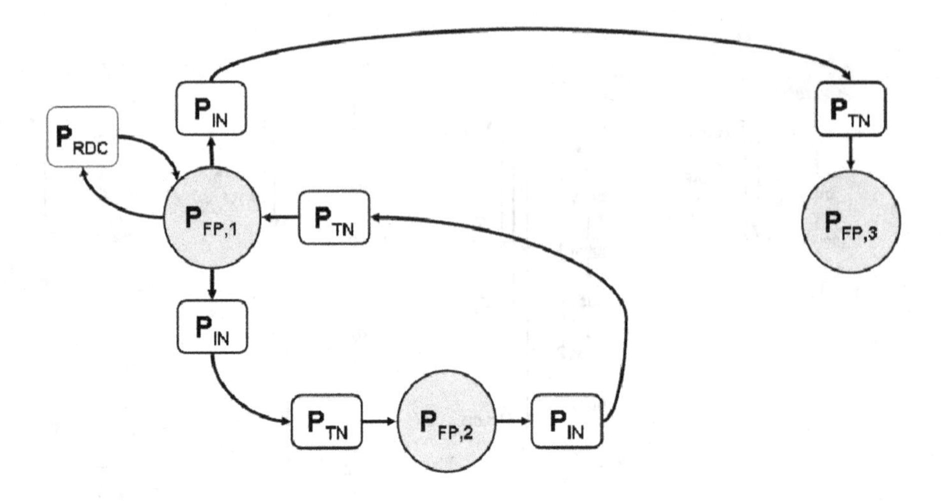

Figure 7.5. Reactive Process Network with Timing Annotation Nodes

projected tags. The latter is calculated by adding the respective delay annotation to the current tag.

DEFINITION 7.12 (INITIATOR NODE) *Given a set of 2 functional signals $\mathcal{S}^2$. An Initiator Node $\mathcal{P}_{IN} \in \mathcal{S}^2$ is a functional process given by the 6-tuple $\mathcal{P}_{IN} = (s_{IN,I}, s_{IN,O}, \mathcal{U}_{IN,\,projected}, \mathcal{D}_{IN}, f_{IN,activate}, f_{IN,update},)$ with*

- *one input signal $s_{IN,I}$,*
- *one output signal $s_{IN,O}$,*
- *a Delay Queue $\mathcal{U}_{IN,projected} \subseteq \mathcal{T} \times \mathcal{V}_{ADT}$ according to definition 7.5, which stores projected events according to the chronological order of the event tags,*
- *a set of delays $\mathcal{D}_{IN} = \{\Delta t_{IN,delay,i}\}, \Delta t_{IN,delay,i} \in \mathbf{N}^+$,*
- *one state transition function $f_{IN,activate}$, which is executed on the arrival of incoming events. $f_{IN,activate}$ first calculates the new tag, inserts the respective event into the list of projected events and eventually schedules the activation of the $f_{IN,update}$ function.*

  $f_{IN,activate}\{$
  // activated on the arrival of new events $e_I = (t_I, v_I) = ((t_{I1}, t_{I2}), v_I)$
  $e_{new} = ((t_{I1} + \Delta t_{IN,delay}, 0), v_I)$;
  $\mathcal{U}_{IN,projected}$.*insert*($e_{new}$);
  if ($\mathcal{U}_{IN,projected}$.*length*() $==1$) { *schedule_update_function*(e_{new}.*t*); }
  }

- *one state transition function $f_{IN,update}$, which is executed when the simulation time reaches the tag of the first event in the Delay Queue. $f_{IN,update}$ removes the first event from the list, forwards it to the output signal, and resumes at the tag of the new head of list event.*

  $f_{IN,update}\{$
  $e_O = \mathcal{U}_{IN,projected}$.*getFirst*();
  if ($\mathcal{U}_{IN,projected}$.*length*() > 0) {

```
    e_head = U_IN,projected.readFirst();
    schedule_update_function(e_head.t);
  }
}
```

Target Node. The *Target Node* is connected to an input signal of a Functional Process and implements the annotation of an initiation interval. In essence, the Target Node *guards* the activation of the process against the arrival of incoming events depending on the current state of the processing element. In case the processing resource is free, the event is immediately forwarded to the attached Functional Process and the node internal state changes to busy. On the other hand, in case the resource is already busy at the arrival of a new event, the incoming event is delayed until the resource is free.

DEFINITION 7.13 (TARGET NODE) *Given a set of 2 functional signals $\mathcal{S}^2$. A Target Node $\mathcal{P}_{TN} \in \mathcal{S}^2$ is a functional process given by the 7-tuple $\mathcal{P}_{TN} = (s_{TN,I}, s_{TN,O}, \mathcal{U}_{TN,pending}, u_{TN,busy}, \mathcal{D}_{TN}, f_{TN,activate}, f_{TN,update})$*

- *one input signal $s_{TN,I}$,*
- *one output signal $s_{TN,O}$,*
- *a FIFO Queue $\mathcal{U}_{TN,pending} \subseteq \mathcal{T} \times \mathcal{V}_{ADT}$ according to definition 7.4 which stores pending events in First-In-First-Out order,*
- *a boolean state variable $u_{TN,busy}$, which is initialized with false,*
- *a set of initiation intervals $\mathcal{D}_{TN} = \{\Delta t_{TN,init,i}\}, \Delta t_{TN,init,i} \in \mathbf{N}^+$*
- *a state transition function $f_{TN,activate}$, which is executed on the arrival of incoming events, and appends the value of the incoming event to the FIFO Queue. In case the processing element is currently not busy, the $f_{TN,update}$ function is called immediately.*

```
f_TN,activate{
  U_TN,pending.enqueue(e_I);
  if (u_TN,busy == false) { schedule_update_function(); } }
```

- *one state transition function $f_{TN,update}$, which is executed when the current initiation interval is ended. $f_{TN,update}$ removes the first event from the list, forwards it to the output signal, calculates the next initiation interval and resumes until the next initiation interval is ended.*

```
f_TN,update{
  if (U_TN,pending.length() == 0) {
    u_TN,busy = false;
  }
  else {
    e_O = U_TN,pending.getFirst();
    t_init = t_now + Δt_TN,init,i
    u_TN,busy = true;
    schedule_update_function(t_init);
  }
}
```

Figure 7.6 depicts an example composition of a single timing annotated Functional Process together with the internal structure of the timing annotation nodes. Looking from the viewpoint of the outer signals $s_{TN,I}$ and $s_{IN,O}$, the timing behavior of the composition corresponds to the execution of the Functional Process on a physical processing element.

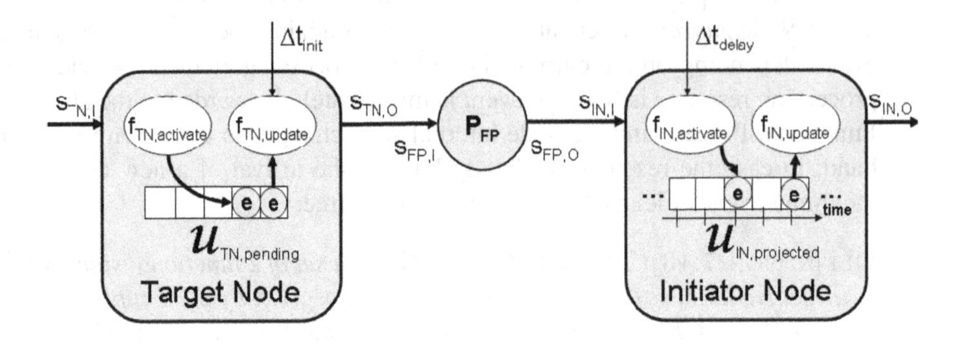

Figure 7.6. Functional Process with Timing Annotation Nodes

The timing annotated Reactive Process Network already captures SoC architectures, which exclusively employ single threaded processing elements and point to point connections. This expressiveness is usually sufficient for modeling the data-plane processing of high performance applications like backbone packet switches [215] or 3D graphic processors [218].

So far, it is not possible to modele the performance of shared processing elements, which cuncurrently execute multiple processes: the timing annotation nodes are restricted to the timing characteristics of a single Functional Process. Instead, interleaved task execution on multi-threaded Virtual Processing Units requires a joint consideration of all processes mapped to a single VPU.

7.3.2 Virtual Processing Unit Timing Model

This section extends the timing model towards the modeling of interleaved execution of functional processes on a Virtual Processing Unit (VPU). VPUs are a generalized representation of multi-threaded processing elements, and capture the notion of both hardware multi-threading (HW-MT) as well as software operating systems (SW-OS). In the considered context of abstract performance modeling, the difference between both implementation choices is merely expressed by the delay penalty of a task swap[5].

[5] Please refer to section 3.2.3 on page 19

In conformity with the timing annotation nodes defined in the previous section, the processing delay characteristics of an individual Functional Process are still specified using delay and initiation interval annotations. Now the VPU timing model provides means to express the timing dependencies caused by task-level resource sharing of processing elements.

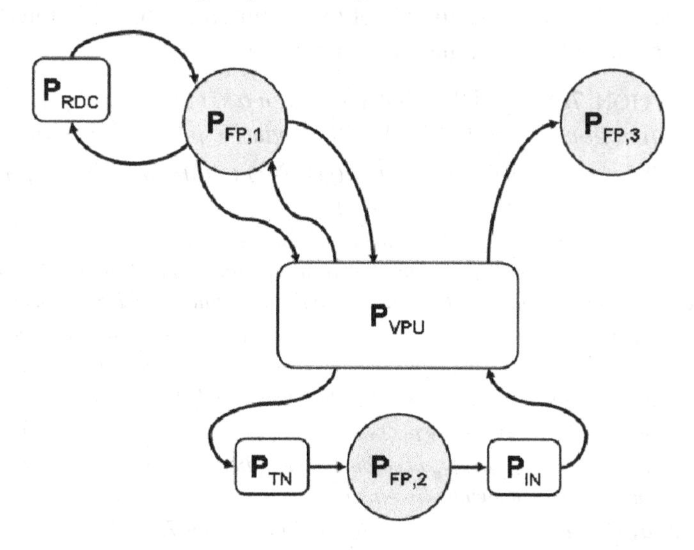

Figure 7.7. Timing Annotated Process Network with $P_{FP,1}$ and $P_{FP,3}$ mapped to a VPU

Similar to the Target and Initiator Nodes, the VPU Node a) guards the activation of the associated Functional Processes against the arrival of new events and b) delays the produced events. In contrast to the individual timing annotation nodes, the VPU Node controls *all* input and output signals of *all* Functional Processes mapped to the considered VPU. As depicted in the example in figure 7.7, the individual Target and Initiator Nodes of the Functional Processes $\mathcal{P}_{FP,1}$ and $\mathcal{P}_{FP,3}$ are removed. Instead the input and output signals of the processes mapped to a VPU are directly connected to the VPU Node $\mathcal{P}_{VPU}$.

In addition to the initiation intervals and processing delays of the individual Functional Processes, the VPU node also takes the delay penalty of a task swapping $\Delta t_{VPU,swap}$ and task preemption $\Delta t_{VPU,preemption}$ into account. The latter captures the fact, that a high priority process might replace a low priority process. In this case the initiation interval of the replacing process has to be added to the processing delay of the replaced process[6]

As a first step from the idealized point-to-point communication of the Reactive Process Network towards physical communication resources, the VPU

[6]Please refer to the example in section 6.2.2 on page 66

Node distinguishes between a *functional interface* and an *communication interface*. The functional VPU interface is complementary to the associated Functional Processes such that every signal of a functional VPU interface $s_{VPU,F}$ is connected one-to-one to a signal of a Functional Process. On the other hand, the communication interface corresponds to the physical ports of the modeled processing element. The signals of the communication VPU interface $s_{VPU,C}$ are connected to the external system context.

DEFINITION 7.14 (VPU NODE) *Given a set of* $n = N_{CI} + N_{CO} + N_{FI} + N_{FO}+$ *functional signals* $\mathcal{S}^n$. *A VPU Node* $\mathcal{P}_{VPU} \in \mathcal{S}^n$ *is a process given by the 8-tuple* $\mathcal{P}_{VPU} = (\mathcal{S}_{VPU,I}, \mathcal{S}_{VPU,O}, \mathcal{E}_{VPU,Internal}, \mathcal{U}_{VPU,pending}, \mathcal{U}_{VPU,projected}, \mathcal{U}_{VPU}, \mathcal{D}_{VPU,swap}, \overline{f}_{VPU})$

- *a set of* $N_I = N_{CI} + N_{FI}$ *input signals* $\mathcal{S}_{VPU,I} = \mathcal{S}_{VPU,CI} \cup \mathcal{S}_{VPU,FI}$, *where* $\mathcal{S}_{VPU,CI}$ *denotes a set of input signals connected to the on-chip communication network and* $\mathcal{S}_{VPU,FI}$ *denotes a set of input signals connected to the associated functional process.*
- *a set of* $N_O = N_{CO} + N_{FO}$ *output signals* $\mathcal{S}_{VPU,O} = \mathcal{S}_{VPU,CO} \cup \mathcal{S}_{VPU,FO}$, *where* $\mathcal{S}_{VPU,CO}$ *denotes a set of output signals connected to the on-chip communication network and* $\mathcal{S}_{VPU,FO}$ *denotes a set of output signals connected to the associated functional process.*
- *a set of two internal events* $\mathcal{E}_{VPU,Internal} = \{ e_{up\_update}, e_{down\_update}\}$
- *a FIFO Queue* $\mathcal{U}_{VPU,pending}$ *according to definition 7.4, which stores event arriving at the communication input signals* $\mathcal{S}_{VPU,CI}$,
- *a Delay Queue* $\mathcal{U}_{VPU,projected}$ *according to definition 7.5, which delays events arriving at the functional input signals* $\mathcal{S}_{VPU,F}$ *to account for annotated delays as well as the for additional delays due to task preemption and swapping,*
- *a set of internal variables* $\mathcal{U}_{VPU} = \{ u_{busy}, u_{prio}, \Delta t_{swap}\}$, *where*
  - *a state variable* u_{busy}, *which is initialized with false*
  - *a variable* u_{prio} *retains the priority of the active process*
- *a set of swapping delays* $\mathcal{D}_{swap} = \{\Delta t_{swap,i}\}, \Delta t_{swap,i} \in \mathbf{N}^+$
- *a set of state transition functions* $\overline{f}_{VPU}$:
  - $f_{VPU,up\_activate}$ *sensitive to events on the signals* $\mathcal{S}_{VPU,CI}$ *connected to the communication network*
  - f_{VPU,up_update} *sensitive to the internal event* e_{up_update}
  - $f_{VPU,down\_activate}$ *sensitive to the events on the signals* $\mathcal{S}_{VPU,FI}$ *connected to the associated functional process*
  - $f_{VPU,down\_update}$ *sensitive to the internal event* e_{down_update}

The internal structure of the VPU Node is depicted in figure 7.8. The operational semantics of the VPU functions are individually defined and explained in the following paragraphs. Recall that according to the packet-level TLM paradigm all values are represented as Abstract Data Types (ADTs). In the context of the VPU mapping, each ADT provides a number of predefined members to store the priority, delay and state of the actual event. The tag and the value fields of the event are accessible by the point operator, i.e. $e_i.v.prio$ denotes the $priority$ field of event e_i.

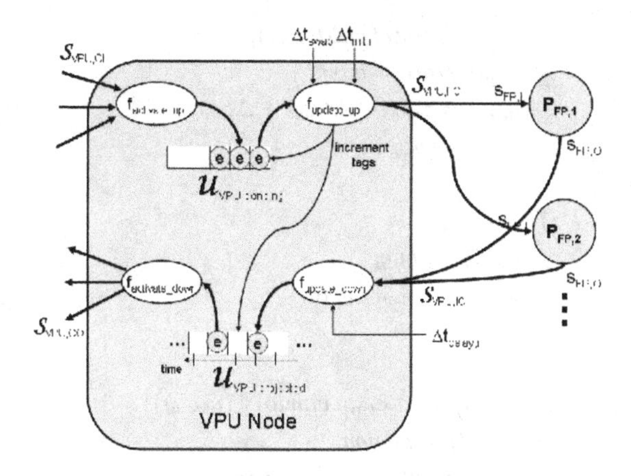

Figure 7.8. VPU Node

Up-Stream Event Processing

The 'up-stream' path refers to the processing of events, which arrive at the communication input signals $\mathcal{S}_{CI}$ and are forwarded to the functional output signals. Similar to the Target Node, new events are first inserted into the $\mathcal{U}_{VPU,pending}$ queue of pending events by the $f_{VPU,up\_activate}$ function. In consideration of the current VPU state, the events are forwarded to the target Functional Process by the f_{VPU,up_update} function.

The $f_{VPU,up\_activate}$ function always inserts the arriving events into the FIFO queue $\mathcal{U}_{VPU,pending}$. In case of preemption or idle state, this method additionally activates f_{VPU,up_update} to update the internal state of the VPU Node.

$f_{VPU,up\_activate}\{$
$\quad \mathcal{U}_{VPU,pending}.insert(e_{VPU,AI});$
$\quad if((u_{prio} < e_{VPU,AI}.v.prio) \,\|\, (!u_{busy})))$
$\qquad \{e_{up\_update}.notify();\}$
$\}$

The function f_{VPU,up_update} handles the tag manipulation of both pending events for activation of the mapped functional processes as well as outgoing events generated by the processes.

0 $\quad f_{up\_update}\{$
1 $\quad\quad remove\_finished\_processes(\mathcal{U}_{VPU,pending});$
2 $\quad\quad if(\mathcal{U}_{VPU,pending}.notEmpty())\{$
3 $\quad\quad\quad if(e_{tmp} = schedule\_process(\mathcal{U}_{VPU,pending}))\{$
4 $\quad\quad\quad\quad if(e_{tmp}.v.state == init)\{$
5 $\quad\quad\quad\quad\quad e_{VPU,FO} = e_{tmp};$

```
6              e_{VPU,FO}.notify(Δt_swap);
7              e_tmp.v.state = busy;
8              e_tmp.t = t_now + Δt_init + Δt_swap;
9              u_init = true;
10           }
11           if(!u_busy){
12              u_busy = true;
13              u_prio = e_tmp.v.prio;
14           }
15           else{
16              if((u_prio < e_tmp.v.prio) || (u_init)){
17                 // preemption
18                 for((all events in U_{VPU,pending})&&(u_busy == true))
19                    event.t += Δt_init + Δt_swap;
20                 for(all events in U_{VPU,projected})
21                    event.t += Δt_init + Δt_swap;
22                 u_prio = e_tmp.v.prio;
23                 u_init = false;
24              }
25              else{ // resume functional process
26                 for((all events in U_{VPU,pending})&& (u_busy == true))
27                    event.t += Δt_swap;
28                 for(all events in U_{VPU,projected})
29                    event.t += Δt_swap;
30                 Δt_init = e_tmp.t − t_now;
31                 u_prio = e_tmp.v.prio;
32              }
33           }    // end if(!u_busy)
34           e_up_update.notify(Δt_init + Δt_swap);
35        }
36     }
37     else{
38        u_busy = false;  u_prio = −1;
39     }    // end if(U_{VPU,pending}.notEmpty())
40  }
```

On every execution of f_{VPU,up_update} with a non-empty priority queue, the VPU checks whether a new functional process needs to be scheduled from the set of pending events in the pending queue $\mathcal{U}_{VPU,pending}$ (line 3). In case an event has been scheduled for the first time (condition in line 4 is true), the delayed activation of the functional task takes the penalty for task swapping into account (line 6).

The functional process is activated and the VPU performs the required timing manipulation of the events. The value Δt_{init} denotes the individual initiation interval according to definition 7.13 of the activated process. At the same time the new events generated during the process activation are inserted into the delay queue $\mathcal{U}_{VPU,projected}$. These events remain inside this queue until their tag is due. As depicted in figure 7.8, sending of the projected events is handled by the functions $f_{VPU,down\_update}$ and $f_{VPU,down\_activate}$.

Task preemption occurs when the current task has a lower priority than the selected task. In this case all generated events in the delay queue $\mathcal{U}_{VPU,projected}$ and tags of already activated processes in the pending queue $\mathcal{U}_{VPU,pending}$ have to be increased (lines 18 – 21). The preemption time is calculated from the initiation interval Δt_{init} of the displacing processes and the required swapping time Δt_{swap}.

The tag incrementation of already generated events in case of task resuming is performed in lines 26 – 29. Finally the next activation of f_{VPU,up_update} occurs after the current task is finished, i.e. after the initiation time of the active process Δt_{init} and the swapping penalty Δt_{swap} (line 34). If no events are pending in the priority queue, the VPU switches to idle state and waits for the arrival of new events (line 38).

Down-Stream Event Processing

The 'down-stream' path forwards events from the functional input signals to the architectural output signals. Before these events are delayed by the Initiator Nodes to satisfy the processing delay of the respective functional process. Now this task is performed by the VPU Node to account for further delays in case of take task swapping and preemption. The individual delay annotation Δt_{delay} of the event arriving at the functional input signals $\mathcal{S}_{FI}$ is accessible as a specific field of the event value $e_{VPU,FI}.v.delay$.

After insertion of new events into the delay queue $\mathcal{U}_{VPU,projected}$ by the $f_{VPU,down\_activate}$ function, the events are further delayed in case of preemption (lines 21/22 and 28/29 of $f_{VPU,up\_activate}$). Naturally the activate and update function in the down-stream closely resemble their counterpart in the Initiator Node definition 7.12.

$f_{VPU,down\_activate}\{$
// activated on new events $e_I = e_{VPU,FI}$
$e_{new} = e_{VPU,FI}$;
$e_{new}.t = t_{now} + e_{VPU,FI}.v.delay$;
$\mathcal{U}_{VPU,projected}$.insert($e_{new}$);
$e_{head} = \mathcal{U}_{VPU,projected}$.readFirst();
}

On expiration of their final deadline, events are dequeued by the $f_{VPU,down\_update}$ function and forwarded to the communication output signals. In case of VPU internal communication, events are immediately inserted into the $\mathcal{U}_{VPU,pending}$ list.

$f_{VPU,down\_update}$\{
// activated when simulation time reaches head of list tag
$e_{VPU,CO} = \mathcal{U}_{VPU,projected}$.getFirst();
$e_{head} = \mathcal{U}_{VPU,projected}$.readFirst();
if ($\mathcal{U}_{IN,projected}$.length() > 0) \{ schedule_update_function($e_{head}.t$); \}
\}

Note, that elegance of the VPU timing scheme is founded on the principle, that events are always projected into the future and do not become effective before the correct point in time. By that, the Functional Processes are always activated at the correct point in time and based on the correct input data. The correctness is ensured by definition 7.7 of CEFSM building blocks, which enforces independence from external events during the execution of a basic block. In other words, preemption influences neither the behavior nor the *individual* timing of any process activation. By that, the VPU timing scheme averts the need for complex and expensive roll-back mechanisms.

7.3.3 Communication Timing Model

Until now, the modeling of the communication architecture is restricted to idealized point-to-point communication. This section extends the timing model with *Communication Nodes*, which capture the impact on performance of physical on-chip communication architectures.

In essence, the total delay Δt_{CN} any type of communication architecture imposes on the transactions can be divided into three components: First, the pending time $\Delta t_{CN,pending}$ is the time spent for waiting until the communication resource is available. $\Delta t_{CN,pending}$ depends on the number and bandwidth of existing resources as well as on the current amount of events competing for the resources. Second, $\Delta t_{CN,next\_arbitrate}$ denotes the duration until the calculation of the grant for the next transaction starts. Last, $\Delta t_{CN,transfer}$ is the time spent for the transfer of the data, which is entirely determined by the size of the data and the bandwidth of the resource.

As depicted in figure 7.9, the internal structure of a Communication Node somewhat resembles a merged Target and Initiator Node. Indeed, the transfer delay $\Delta t_{CN,transfer}$ corresponds to the processing delay parameter $\Delta t_{IN,delay}$ and $\Delta t_{CN,next\_arbitrate}$ corresponds to the initiation interval $\Delta t_{TN,init}$. As a major difference, the Communication Nodes issues two output events for every incoming event, since the beginning as well as the end of a transfer may be of importance for the consuming process.

The close analogy of the Communication Node and the Target/Initiator Nodes is of course little surprising: On this high level of abstraction, the impact on

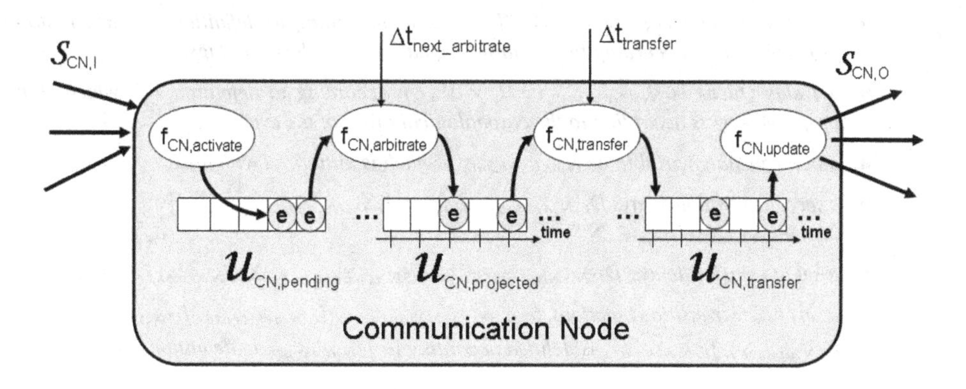

Figure 7.9. Communication Node

performance of a communication node is not different from a processing element. The definition of a dedicated Communication Node is justified by the observation, that the impact of a certain type of communication architecture is captured by a specific arbitration function.

As discussed later in chapter 8, this *factorization* enables the creation of a unified framework, which can be efficiently parameterized to model any kind of communication architecture [24]. Despite the simplicity of the basic model, the implementation of the arbitration function $f_{CN,arbitrate}$ and the calculation of the actual timing parameters $\Delta t_{CN,next\_arbitrate,i}$ and $\Delta t_{CN,transfer,i}$ can become very demanding for complex real-world communication architectures.

According to the discussion in section 6.1.3, both start-of-transfer (sot) event and end-of-transfer (eot) event are important in the context of packet-level data modeling. Hence both events are generated by the $f_{CN,update}$ function. The actual type of event is marked in the current value $v = e.v \in \mathcal{V}_{ADT}$ as a dedicated field $v.event\_type$ of the Abstract Data Type. Additionally, in general more than one consumer process can be connected to the output signals of the Communication Node. The target process is identified by an additional field $v.target$ of the ADT.

DEFINITION 7.15 (COMMUNICATION NODE) *Given a set of* $n = N_I + N_O$ *functional signals* $\mathcal{S}^n$*. A Communication Node* $\mathcal{P}_{CN} \in \mathcal{S}^n$ *is a process given by the 9-tuple.* $\mathcal{P}_{CN} = (S_{CN,I}, S_{CN,O}, \mathcal{U}_{CN,pending}, \mathcal{U}_{CN,projected}, \mathcal{U}_{CN,transfer}, u_{CN,arbiter\_busy}, \mathcal{D}_{CN,next\_arbitrate}, \mathcal{D}_{CN,transfer}, \bar{f}_{CN})$

- *a set of input signals* $S_{CN,I}$,
- *a set of output signal* $S_{CN,O}$,
- *a FIFO queue* $\mathcal{U}_{CN,pending} \subseteq \mathcal{T} \times \mathcal{V}_{ADT}$ *according to definition 7.4, which stores pending events in First-In-First-Out order,*

- *a Delay Queue* $\mathcal{U}_{CN,projected} \subseteq \mathcal{T} \times \mathcal{V}_{ADT}$ *according to definition 7.5, which stores projected events according to the chronological order of the event tags,*
- *a Delay Queue* $\mathcal{U}_{CN,transfer} \subseteq \mathcal{T} \times \mathcal{V}_{ADT}$ *according to definition 7.5, which stores projected events according to the chronological order of the event tags,*
- *a boolean state variable* $u_{CN,arbiter\_busy}$, *which is initialized with false,*
- *a set of arbitration delays* $\mathcal{D}_{CN,next\_arbitrate} = \{\Delta t_{CN,next\_arbitrate,i}\}$, $\Delta t_{CN,next\_arbitrate,i} \in \mathbf{N}^+$,
- *a set of transfer delays* $\mathcal{D}_{CN,transfer} = \{\Delta t_{CN,transfer,i}\}$, $\Delta t_{CN,transfer,i} \in \mathbf{N}^+$,
- *a set of state transition functions* $\bar{f}_{CN} = (f_{CN,activate}, f_{CN,arbitrate}, f_{CN,transfer}, f_{CN,update})$. $f_{CN,activate}$ *is defined according to* $f_{\Gamma N,activate}$ *in definition 7.13:*

  ```
  fCN,activate{
    // activated on arrival of new event eI on one of the input signals SCN,I
    UCN,projected.enqueue(eI);
    if (uCN,arbiter_busy == false) { schedule_arbitrate_function(); }
  }
  ```

  $f_{CN,transfer}$ *issues start-of-transfer (sot) events*[7], *calculates end-of-transfer (eot) event tags and inserts events into the transfer queue*

  ```
  fCN,transfer{
      // activated when simulation time reaches head of list tag
      while (UCN,projected.readFirst().t == tnow) {
          etmp = UCN,projected.getFirst();
          etmp.v.event_type = sot;
          eO[etmp.v.target] = etmp; // notify sot event
          etmp.t = tnow + Δttransfer // calculate eot tag
          UCN,transfer.insert(etmp);
      } // end while
  }
  ```

  $f_{CN,update}$ *issues end-of-transfer (eot) events*[8]*:*

  ```
  fCN,update{
      // activated when simulation time reaches head of list tag
      while (UCN,transfer.readFirst().t == tnow) {
          etmp = UCN,projected.getFirst();
          etmp.v.event_type = eot;
          eO[etmp.v.target] = etmp; // notify eot event
      } // end while
  }
  ```

As defined below, $f_{CN,arbitrate}$ calculates the timing parameters and is therefore specific for the type of the communication architecture.

[7] multiple events can be issued simultaneously in case of parallel communication resources

[8] multiple events can be issued simultaneously in case of parallel communication resources

In the following the respective specialization of the arbitration function for point-to-point link, bus and crossbar communication is defined together with the calculation of timing parameters. In the examples, the size of the current value $v = e.v \in \mathcal{V}_{ADT}$ is accessible as a dedicated field $v.size$ of the Abstract Data Type. Note, that the following selection of timing calculation functions is by far not exhaustive and merely captures the basic types of communication architectures, but the modular structure enables efficient creation and modification of further communication nodes.

Link Node

The point-to-point *Link Node* models transactions between a single source and a single destination module over an exclusive connection. Because of the FIFO semantics of the pending queue $\mathcal{U}_{CN,pending}$, this Link Node effectively corresponds to a FIFO channel. The available bandwidth of the FIFO connection is configured by means of a *bit_width* and a *clock_period* parameter.

DEFINITION 7.16 (LINK NODE) *A Link Node $\mathcal{P}_{Link}$ is a specialization of the Communication Node according to definition 7.15 with $N_I = N_O = 1$.*

$f_{Link,arbitrate}${
  // activated when new event arrived or next_arbitrate interval is ended
  if ($\mathcal{U}_{CN,pending}$.*length()* $==0$) {
    $u_{CN,arbiter\_busy}$ = *false;*
  }
  else {
    $e_{tmp} = \mathcal{U}_{CN,pending}$.*getFirst();*
    $\Delta t_{CN,transfer} = \lceil e_{tmp}.v.size/bitwidth \rceil * clock\_period;$
    $e_{tmp} = ((t_{now}, 0), v_{tmp});$
    $\mathcal{U}_{CN,projected}$.*insert*($e_{tmp}$);
    $t_{next\_arbitrate} = t_{now} + \Delta t_{CN,transfer}$
    $u_{CN,arbiter\_busy}$ = *true;*
    schedule_arbitrate_function($t_{next\_arbitrate}$);
  }
}

Bus Node

The *Bus Node* models data exchange over a shared bus medium. This node can be further specialized towards priority or TDMA like bus arbitration schemes by different implementations of the *select()* algorithm, which determines the grant from the list of pending requests. The priority based bus arbiter simply selects the request with highest priority, whereas the TDMA arbiter is based on a static allocation table.

Similar to the Link Node, the Bus Node is configured by means of a *bit_width* and a *clock_period* parameter to determine the duration of the transaction. Additionally, the calculation of the next grant may be inflicted with an arbitration penalty $\Delta t_{arbitration}$. The example below models a pipelined bus architecture i.e. the arbitration is performed in parallel with the ongoing transfer to hide the arbitration delay.

DEFINITION 7.17 (BUS NODE) *A Bus Node $\mathcal{P}_{Bus}$ is a specialization of the Communication Node according to definition 7.15.*

$f_{Bus,arbitrate}${
  // activated when new event arrived or next_arbitrate interval is ended
  if ($\mathcal{U}_{CN,pending}$.length() == 0) {
    $u_{CN,arbiter\_busy}$ = false;
  }
  else {
    $e_{tmp} = \mathcal{U}_{CN,pending}$.select();
    $\Delta t_{CN,transfer} = \lceil e_{tmp}.v.size/bitwidth \rceil * clock\_period;$
    $t_{start\_of\_transfer} = t_{now} + \Delta t_{CN,arbitrate}$
    $e_{tmp} = ((t_{start\_of\_transfer}, 0), v_{tmp});$
    $\mathcal{U}_{CN,projected}$.insert($e_{tmp}$);
    $t_{next\_arbitrate} = t_{now} + \Delta t_{CN,transfer} - \Delta t_{CN,arbitrate}$
    $u_{CN,arbiter\_busy}$ = true;
    schedule_arbitrate_function($t_{next\_arbitrate}$);
  }
}

Crossbar Node

The *Crossbar Node* models on-chip network architectures, which are able to perform more than one transaction at the same time. Similar to the Bus Node, the Crossbar Node can be specialized to different arbitration algorithms like e.g. static, weighted and un-weighted arbitration by implementing of the `select()` method accordingly. The Crossbar Node maintains an internal $N_O \times P$ data structure $\mathcal{U}_{fifo}$, which queues requests per destination module and per priority in a FIFO queue according to definition 7.4. For every arbitration interval, the `select()` algorithm generates a $N_O \times P$ grant matrix G from the head-of-line $N_O \times P$ snapshot matrix S of the queues. Since the packet size is assumed to be fixed, $\Delta t_{CN,transfer}$ in this case is a constant parameter.

DEFINITION 7.18 (CROSSBAR NODE) *A Crossbar Node $\mathcal{P}_{Crossbar}$ is a specialization of the Communication Node according to definition 7.15.*

$f_{Crossbar,arbitrate}${
  // activated when new event arrived or next_arbitrate interval is ended
  if ($\mathcal{U}_{CN,pending}$.length() == 0) {
    $u_{CN,arbiter\_busy}$ = false;
  }
  else {
    while ($\mathcal{U}_{CN,pending}$.length() > 0) {
      $e_{tmp} = \mathcal{U}_{CN,pending}$.dequeue();
      $\mathcal{U}_{fifo}[e_{tmp}.v.target][e_{tmp}.v.prio]$.enqueue($e_{tmp}$);

```
        }
        t_start_of_transfer = t_now + Δt_CN,arbitrate
        S = generate_hol_snapshot(U_fifo);
        G = select(S);
        for (i = 0; i < N_O; i++) {
            for (j = 0; j < P; j++); {
                if (G[i][j] == true) {
                    e_tmp = U_fifo[i][j].dequeue();
                    e_tmp = ( (t_start_of_transfer, 0), v_tmp);
                    U_CN,projected.insert(e_tmp);
                } // end if
            } // end for j
        } // end for i
        t_next_arbitrate = t_now + Δt_CN,transfer − Δt_CN,arbitrate
        u_CN,arbiter_busy = true;
        schedule_arbitrate_function(t_next_arbitrate);
    }
}
```

7.3.4 Architecture Process Network

Now all necessary ingredients are available to construct an architecture model in terms of an Architecture Process Network as depicted in figure 7.10.

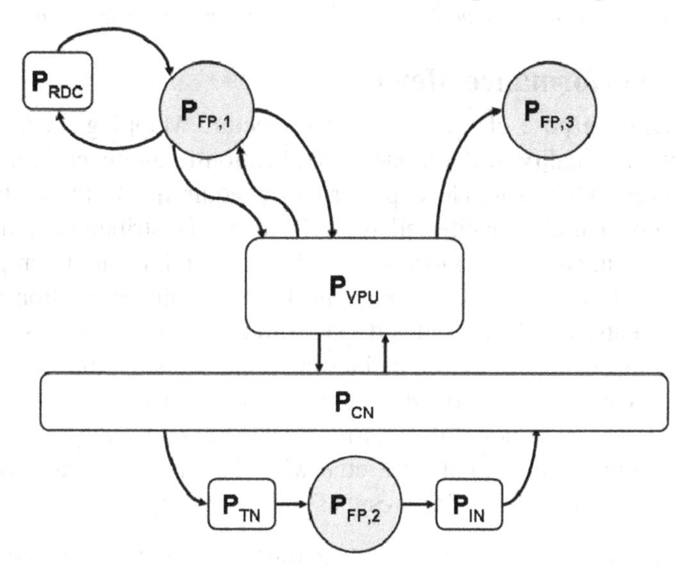

Figure 7.10. Architecture Process Network

An Architecture Process Network is successively constructed from a Reactive Process Network according to definition 7.11 by inserting Initiator and Target Nodes for the mapping onto single-threaded processing elements, inserting VPU Nodes onto multi-threaded processing elements, and inserting Network

Nodes to model the communication architecture. Intuitively, the insertion of the timing annotation nodes *hooks* the functional events to the architectural resources.

DEFINITION 7.19 (ARCHITECTURE PROCESS NETWORK) *An Architecture Process Network is a 7-tuple* $\mathcal{N}_{AP} = (\bar{\mathcal{P}}_{FP}, \bar{\mathcal{P}}_A, \bar{\mathcal{P}}_{PC}, \mathcal{M}_{PE,I}, \mathcal{M}_{PE,O}, \mathcal{M}_{CN,I}, \mathcal{M}_{CN,O})$, *where*

- $\bar{\mathcal{P}}_{FP}$ *is set of Functional Processes according to definition 7.8.*
- $\bar{\mathcal{P}}_A$ *is set of Architecture Nodes* $\mathcal{P}_A \in \{\mathcal{P}_{TN}, \mathcal{P}_{IN}, \mathcal{P}_{VPU}, \mathcal{P}_{CN}\}$,
- $\mathcal{M}_{PE,I}$ *is a bijective mapping, that assigns the input signals* $s \in \mathcal{S}_{FP,I}$ *of each Functional Process* $\mathcal{P}_{FP} \in \bar{\mathcal{P}}_{FP}$ *either to the output signal* $s_{TN,O}$ *of a Target Node* $\mathcal{P}_{TN}$ *or to one functional output signal* $s \in \mathcal{S}_{VPU,FO}$ *of a VPU Node* $\mathcal{P}_{VPU}$
- $\mathcal{M}_{PE,O}$ *is a bijective mapping, that assigns the output signals* $s \in \mathcal{S}_{FP,O}$ *of each Functional Process* $\mathcal{P}_{FP} \in \bar{\mathcal{P}}_{FP}$ *either to the input signal* $s_{IN,I}$ *of an Initiator Node* $\mathcal{P}_{IN}$ *or to one functional input signal* $s \in \mathcal{S}_{VPU,FI}$ *of a VPU Node* $\mathcal{P}_{VPU}$
- $\mathcal{M}_{CN,I}$ *is a bijective mapping, that assigns the input signals* $s \in \mathcal{P}_{CN,I}$ *of each Commmunication Node* $\mathcal{P}_{CN} \in \bar{\mathcal{P}}_A$ *either to the output signal* $s_{IN,O}$ *of an Initiator Node* $\mathcal{P}_{IN}$ *or to one communication output signal* $s \in \mathcal{S}_{VPU,CO}$ *of a VPU Node* $\mathcal{P}_{VPU}$
- $\mathcal{M}_{CN,O}$ *is a bijective mapping, that assigns the output signals* $s \in \mathcal{P}_{CN,O}$ *of each Commmunication Node* $\mathcal{P}_{CN} \in \bar{\mathcal{P}}_A$ *either to the input signal* $s_{TN,O}$ *of an Target Node* $\mathcal{P}_{TN}$ *or to one communication input signal* $s \in \mathcal{S}_{VPU,CI}$ *of a VPU Node* $\mathcal{P}_{VPU}$

7.4 Performance Metrics

The major purpose of the Virtual Architecture Mapping methodology is to evaluate the quality of the modeled application-to-architecture mapping. Besides meeting the system level performance requirements, the quality of a mapping decision is also manifested in a well balanced distribution of the processing and communication workload imposed on the architectural components. The system architect has to avoid bottlenecks to sustain application performance requirements. On the other hand, poor utilization has a needless negative impact on the cost. This section defines the principal performance and utilization metrics, which can be derived from the timing model.

Naturally, the values of the metrics have to be statistically aggregated during the execution of the simulation, otherwise the amount of data would prevent from any interpretation of the results.

DEFINITION 7.20 (STATISTIC AGGREGATION) *The statistic aggregation* $\tilde{M}$ *of a Metric* M *with* N *Values* $\mathcal{M} = \{m_1, \ldots, m_N\}$ *is a 4-tuple* $\tilde{M} = (M_{min}, M_{max}, M_{avg}, h_M)$ *with*

- *the minimum value* $M_{min} = \min_{m_i \in \mathcal{M}}(m_i)$,
- *the maximum value* $M_{max} = \max_{m_i \in \mathcal{M}}(m_i)$,
- *the average value* $M_{avg} = \frac{\sum_{m_i \in \mathcal{M}} m_i}{N}$,

- *the frequency distribution* $f_M : (M_{min}, \ldots, M_{max}) \mapsto \mathbf{N}^+$ *with* $h_M(m_j) = \|\{m_i \in \mathcal{M} \mid m_i = m_j\}\|$

The following subsections define the metrics, which are derived from the annotation nodes, the VPU node and the communication nodes respectively. Additionally, a number of aggregated system level metrics are collected across multiple timing nodes. In all definitions t_{total} denotes the total simulation time and N denotes the total number of events on the considered signal.

7.4.1 Timing Annotation Metrics

Already the statistic aggregation of the atomic Initiator and Target Node timing parameters $\tilde{\Delta} t_{delay}$ and $\tilde{\Delta} t_{init}$ provide insight to the performance of the considered Functional Process and the required timing budget. Recalling that the initiation interval represents the busy periods of a processing element, the utilization can be derived from the accumulated busy periods:

DEFINITION 7.21 (PROCESSING ELEMENT UTILIZATION) *The utilization* U_{PE} *of a single-threaded processing element denotes the ratio of the summarized initiation intervals and the total simulation time:*

$$U_{PE} = \frac{\sum_{n=1}^{N} \Delta t_{init,n}}{t_{total}}$$

In addition to the above performance and utilization metrics, the Target Node enables the detection of bottlenecks. In case the capacity of the anticipated processing element is not sufficient, arriving events are not consumed and therefore stored in the FIFO queue. Hence, the number of pending events $n_{PE,pending}(t_1)$ waiting in the FIFO queue at a certain point in time t_1 reflects the current load situation.

DEFINITION 7.22 (PROCESSING ELEMENT PENDING QUEUE LENGTH) *The Pending Queue Length* $n_{PE,pending}(t_1)$ *denotes the number of projected events at time* $t = t_1$*:*

$$n_{pending}(t_1) = \mathcal{U}_{TN,projected}.length()$$

Correlated to the number of pending events, the consideration of the pending time of events is a significant metric for the identification of bottlenecks.

DEFINITION 7.23 (PROCESSING ELEMENT PENDING TIME) *The pending time of an event e denotes the difference between the tags of the event arrival* $e_{TN,I}.t$ *at the input signal* $s_{TN,I}$ *and the tag of the respective event at the output signal:*

$$\Delta t_{PE,pending}(e) = e_{TN,O}.t - e_{TN,I}.t$$

Obviously, high average values in the statistic aggregation(s) $\tilde{n}_{PE,pending}$ and/or $\tilde{\Delta t}_{PE,pending}$ indicate a permanent bottleneck situation. In consequence, the system architect might either tighten the processing budged allocated for consuming process or relax the budget of the associated producing process(es).

7.4.2 Virtual Processing Unit Metrics

In addition to the individual timing annotation metrics introduced in the previous section, the following VPU metrics enable the investigation of shared processing resources. Again, the utilization and the bottleneck situation are of major interest for the system architect to evaluate the considered application-to-platform mapping.

DEFINITION 7.24 (VPU UTILIZATION) *The VPU utilization U_{VPU} denotes the ratio of the sum of all process busy times and the overall simulation time:*

$$U_{VPU} = \sum_{p=1}^{P} U_{PE,p} = \frac{\sum_{p=1}^{P} \sum_{n=1}^{N_p} \Delta t_{init,n}}{t_{total}}$$

where P denotes the number of processes mapped to the VPU and N_p denotes the number of activations of process p.

In analogy with definition 7.23, the VPU pending time constitutes a significant metric for a bottleneck situation in the shared processing element.

DEFINITION 7.25 (VPU PENDING TIME) *The VPU Pending Time $\Delta t_{VPU,pending}$ of an event e denotes the difference between the tags of the event arrival $e_{VPU,CI}.t$ at the communication input signal $s_{VPU,CI}$ and the tag of the respective event at the functional output signal $s_{VPU,FO}$:*

$$\Delta t_{VPU,pending}(e) = e_{VPU,CI}.t - e_{VPU,FO}.t$$

In order to identify the optimal scheduling algorithm and priority scheme for a considered processing element, the impact of preemption needs to be investigated.

DEFINITION 7.26 (VPU PREEMPTION DELAY) *The VPU Preemption Delay $\Delta t_{VPU,preemption}$ denotes the difference between the original and the effective due time of an event arriving at a functional input signal $s_{VPU,FI}$:*

$$\Delta t_{VPU,preemption} = e_{VPU,CO}.t - (e_{VPU,FI}.t + e_{VPU,FI}.v.delay)$$

In order to efficiently evaluate the scheduling algorithm, the individual preemption delays are aggregated for activating events of a specific task or for all tasks of a specific VPU.

7.4.3 Communication Architecture Metrics

The performance of any type of communication architecture is primarily measured in terms of *throughput* and *latency*.

DEFINITION 7.27 (THROUGHPUT) *The maximum throughput $T_{CN,max}$ is statically determined by the configuration of the communication resource:*

$$T_{CN,max} = \frac{bitwidth}{clock\_period}$$

This upper bound is hardly reached in real systems. Instead the system architecture is often designed to keep the utilization of the communication resources significantly below 100% to minimize contention. Otherwise the real-time behavior of the application cannot be guaranteed. As discussed before, this over-design can only be avoided by quality of service capabilities in the communication nodes.

DEFINITION 7.28 (COMMUNICATION LATENCY) *The communication latency l_{CN} of a single transaction corresponds to the delay between starting and completion events:*

$$l_{CN} = e_{CN,O}.t - e_{CN,I}.t = \Delta t_{CN,pending} + \Delta t_{CN,arbitrate} + \Delta t_{CN,transfer}$$

where in analogy to definition 7.23 $\Delta t_{CN,pending}$ denotes the time the request resides in the pending queue $\mathcal{U}_{CN,pending}$.

The individual latency values can be aggregated with respect to different parameters depending on the current subject of investigation:

- **Node Latency** L_n denotes the aggregated latency values of all transactions on a single communication node:
  $L_n = \tilde{l}_{CN,n}, with\, l_{CN,n} = \{l_{CN} | node = n\}$
- **Quality of Service** $L_{n,p}$ denotes the aggregated latency for transactions with a specific priority value:
  $L_{n,p} = \tilde{l}_{t,n,p}, with\, l_{t,n,p} = \{l_{CN} | node = n \wedge prio = p\}$

The comparison of Quality of Service related latency $L_{n,p}$ metrics highlights the ability of a specific node to separate different service classes. This is of particular interest in the context of more complex on-chip networks.

7.4.4 Application Metrics

The primary goal of system level design is to meet the overall performance requirements of the application. For this purpose, the system architect usually monitors a concise set of system level performance metrics to evaluate the quality of the platform architecture and application mapping. This system level metrics can be considered as either a selection or a composition of the architecture level performance metrics defined in the previous sections.

The *Application Throughput* usually corresponds to the number of events per time unit at the system boundaries. For example the throughput of the Internet Protocol packet processor is measured in terms of number of forwarded IP packets.

The *Application Latency* corresponds to the concatenation of a set of architecture level latencies, which taken together define a *path* of the application events through the system architecture. In analogy to the transistor level world, the *critical path* determines the end-to-end application latency.

7.5 Summary

As a major contribution of this book, this section defines a unified timing model for architecture exploration and task partitioning of arbitrarily complex multi-processor platforms.

Application tasks are represented in terms of timed Communicating Extended Finite State Machines, which capture the coarse-grain functionality, the timing characteristics and the task level parallelism of the application. The application level events exposed by the timing annotated CEFSMs are refined to architecture level events by a set of timing manipulation nodes for processing delay annotation, multi-threaded processing elements and on-chip communication.

Since the mapping onto the timing manipulation nodes is transparent for the application tasks, this timing model enables the rapid exploration of the anticipated architecture in the context of the given application.

Chapter 8

MP-SOC SIMULATION FRAMEWORK

After the introduction of the major concepts and the formal derivation of the underlying timing model in the previous chapters, the discussion now turns to the tool related aspects of MP-SoC platform modeling and exploration. The following sections individually treat on-chip communication modeling, virtual processing units and associated visualization tools.

8.1 The Generic Synchronization Protocol

As already motivated in section 6.1.3, a major feature of the NoC framework is the orthogonalization of the communication services and the communication architecture. By that all user modules communicate through the use of a unified protocol, which is agnostic of any architectural aspects. In particular, the protocol is independent of

- the interface specification, e.g. a specific bus protocol,
- the interconnection scheme, i.e. the protocol is not biased towards point-to-point, bus, or router based communication nodes,
- the topology of the communication network.

The term *synchronization* interface indicates, that the protocol provides merely the basic transport mechanism by means of essential communication primitives for sending and receiving data. Advanced protocol features and network interfaces have to be implemented by higher protocol layers. Here the synchronization interfaces serves as a tunnel for control information, that is exchanged between the user modules and the communication network. In this case, the generality of the interface is of course lost and the design space exploration is limited to a specific communication architecture.

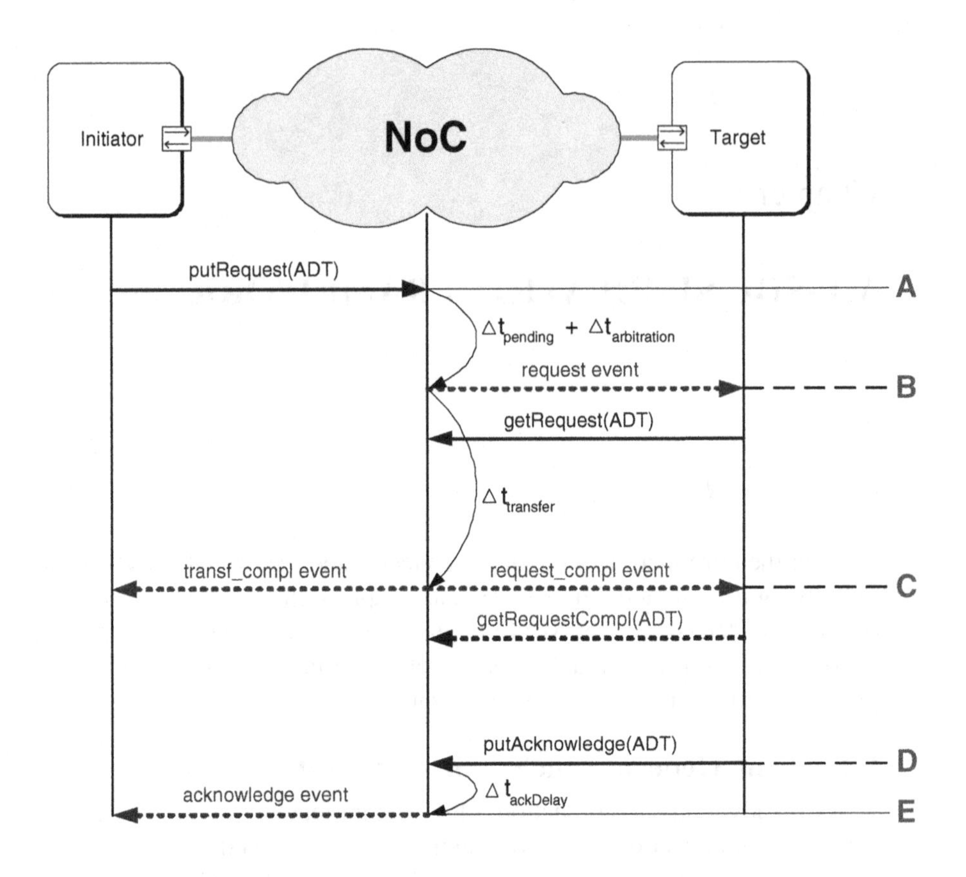

Figure 8.1. Generic Synchronization Interface

The synchronization protocol follows the Open Core Protocol (OCP) [22] semantics, which isolate the interconnect from the interface. However, the existing OCP interfaces at the TL1 and TL2 abstraction are too OCP protocol specific and the timing and data representation is too detailed for the purpose of generic architectural modeling [149, 232]. In fact the concepts of the generic synchronization interface as outlined below have been incorporated into the newly defined OCP TL3 API. Please refer to appendix B on page 163 for an overview of the new TL3 API.

The generic synchronization interface used in the MP-SoC framework is tailored to the packet-level modeling requirements and features timing annotation [233].

As depicted in figure 8.1, the generic synchronization interface offers a confined set of communication methods (full line arrows) and events (dashed arrows). Modules issuing requests are called initiators and those receiving requests are called targets. Transactions are initiated and received by calling these methods and being sensitive to the provided events. The sequence of method calls and event notifications is depicted in the message sequence chart in figure 8.1.

putRequest() ***method*** This method inserts requests into the NoC framework and launches the transaction process.

request ***event*** This event occurs when the NoC Channel starts delivering the packet to the destination target, i.e. the point in time when the first data word arrives at the target.

putRequest() ***method*** This method retrieves the complete packet from the channel.

request_complete ***event*** This target side event occurs at the end of a packet transmission, i.e. the point in time when the last data word arrives at the target.

transf_complete ***event*** This initiator side event occurs at the end of a packet transmission, i.e. the point in time when the last bit arrives at the target.

putAcknowledge() ***method*** This method initiates an acknowledge that is passed back to the initiator. Acknowledge handling is an optional feature.

acknowledge ***event*** This event indicates that the target has send an acknowledge.

Split Transaction Scheme

So far the generic synchronization protocol provides only a unidirectional transportation mechanism. However, transactions like memory reads require a bidirectional information exchange, i.e. the transaction is split in two phases, a request phase and a response phase. Following the OCP specification, split transactions are carried via two independent and symmetric request and response channels. During the response phase, the communicating partners switch roles: the target becomes the initiator and the initiator becomes the target.

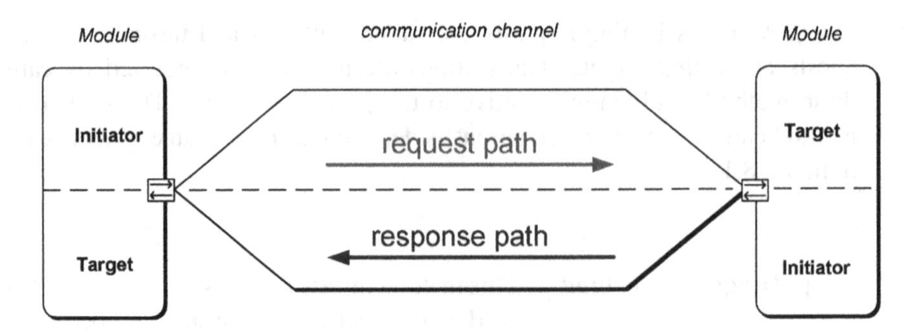

Figure 8.2. NoC Request - Response Path

As depicted in figure 8.2, the NoC communication channel is fundamentally split into independent request and response handling, which makes it adaptable to various split and non-split protocols [232].

Event Chronology

Although the cycle-level handshaking is not seen at the abstraction level of the synchronization interface, the packet-level events can be associated with certain phases in a transaction over a cycle accurate interface. As highlighted in figure 8.3, this relation is the key to the creation of near cycle accurate communication models at the packet abstraction level.

The developer of a packet-level communication model has to match the abstract methods and events with the corresponding transfer[1] in the specific communication interface. In case the considered target architecture naturally adheres to a packet-level interface specification, this approach allows the creation of 100% accurate simulation models. Examples are the STbus architecture and on-chip network interfaces. On the other hand, the rich feature set of processor core specific bus architectures like AMBA AHB or CoreConnect limits the accuracy of packet-level models.

During the simulation, the described events and method calls of figure 8.1 are mapped to discrete simulation time. The markers A, B, C, D and E correspond to this points in the signal trace of figure 8.3.

- **A** The putRequest() method is called. This indicates the point in time when a new request is inserted into the network and the initiator module starts to pass the data to the network. Note that this is not necessarily the time when the transfer starts.

[1] transfer: specific phase in a transaction

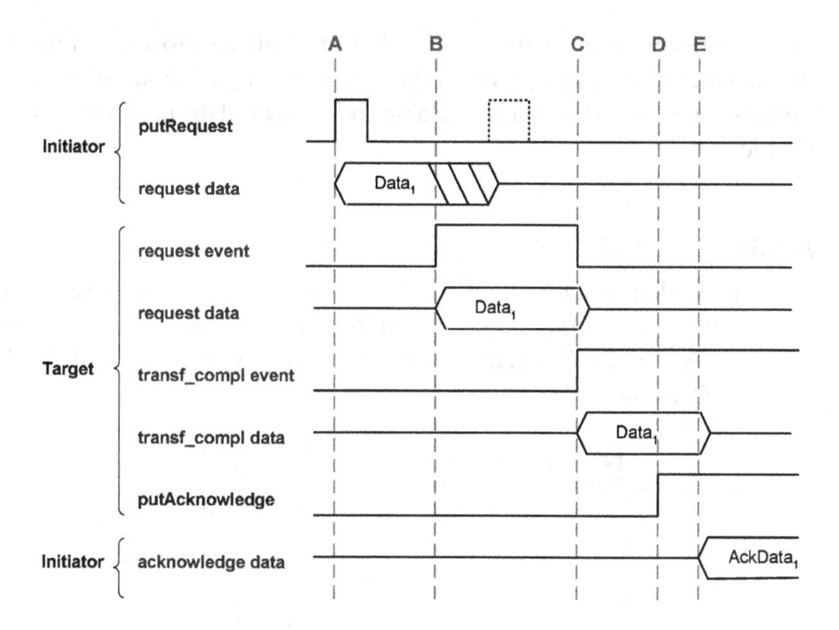

Figure 8.3. Wavetrace visualization of Event Chronology

B The transfer starts when the communication resource is available and the target is ready to accept a new request. When the first word of the packet arrives at the target module the request_event event is notified. The target module can already retrieve the token although the transfer process has not completed. This feature allows the modeling of wormhole and cut-through routing strategies[2].

C At this point in time the last word has been transfered to the target module. This causes the notification of both the request_compl event at the target side as well as the transf_compl event at the initiator side.

D The target module has finished some computation tasks and sends an acknowledge back to the initiator.

E The acknowledge event arrives at the initiator after the time that has been afflicted to the putAcknowledge() call in *D*. The transfer of acknowledges does not claim resources.

A fully non-ambiguous mapping between wavetrace visualization and packet level event model is not possible. The events in figure 8.1 indicate that the data

[2]Please refer to section 3.3.2.0 on page 28

can be retrieved by the activated model. The whole data token is either fetched through the getRequest( )pre-sampling method or getRequestCompl( )post-sampling method. The data is available and valid until the next token overwrites the old one.

Feedback Control

Accurate modeling of the initiator behavior eventually requires to react to events that occur during an ongoing transaction, like for example grant or complete transfers. As depicted in figure 8.4 the NoC framework supports the following different feedback schemes:

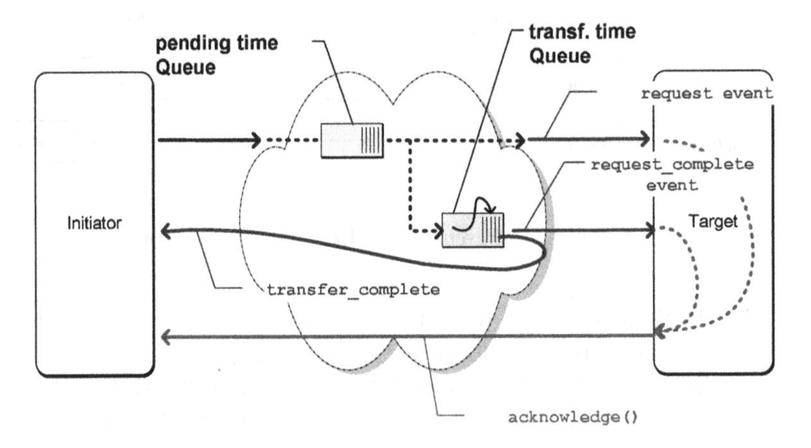

Figure 8.4. NoC Feedback Loops

- **posted request** the initiator does not care about communication status messages. There is no feedback loop.
- **automatic complete** the feedback is automatically generated by the NoC framework. In this case the initiator module is sensitive to the transfer_complete event.
- **user dependent acknowledge** the target module is responsible for acknowledging the initiator. This can be done immediately after the request has arrived or after some processing time. Furthermore it is possible to activate target processing at the begin or the end of the request token.

The flexible selection of the feedback scheme allows the user to trade higher modeling efficiency and simulation speed against higher accuracy.

By following the Open Core Protocol semantics the generic synchronization protocol fully complies to the *Interface Design* principle [70], which advocates the separation of the behavior and communication. This architecture agnostic modeling of the application enables the transparent mapping of the application tasks to architectural elements. The following sections separately introduce the modeling framework for the task mapping to shared processing elements and communication fabrics respectively.

8.2 Generic VPU Model

As outlined in section 6.2.2, the concept of a Virtual Processing Unit captures the notion of shared processing elements like Hardware multi-threading and Software Operating Systems. In that the VPU serves as a task scheduling layer situated between the functional tasks and the communication architecture. In essence the generic VPU model implements the operational semantics of the VPU timing model as derived in section 7.3.2 of the previous chapter. This section gives an overview of the Software architecture, which enables the flexible specification of the task mapping and a modular integration of user defined task scheduling algorithms.

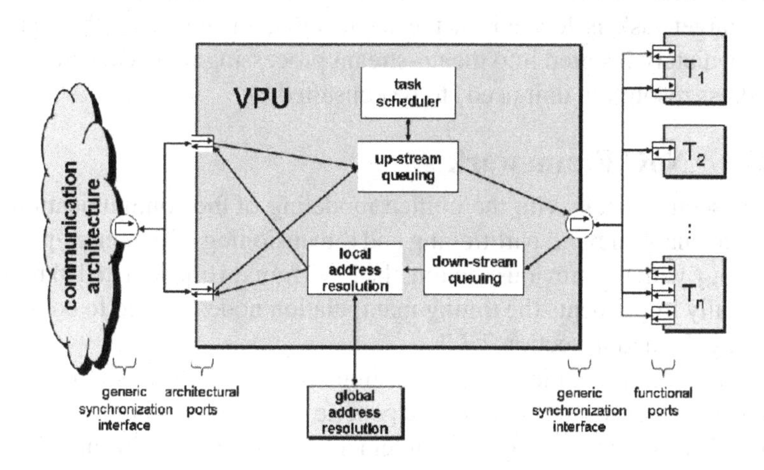

Figure 8.5. VPU Framework Overview

The overall structure of the VPU framework is depicted in figure 8.5. The right hand side shows a number of application tasks, whose functional ports are connected to the generic synchronization interface of a VPU instance. These tasks share a single processing element and a number of architectural ports. The architectural ports are depicted on the left hand side and determine the available bandwidth of the processing elements to the communication network.

The internal structure of the VPU is derived from the VPU Node according to definition 7.14. Here the up-stream queuing component stores incoming events and controls the activation of the application tasks. This comprises the $\mathcal{U}_{VPU,pending}$ queue and the associated handling functions $f_{VPU,up\_activate}$ and f_{VPU,up_update}.

As a separate module of the up-stream processing functionality the VPU contains a task scheduler, which implements the $schedule\_process(\mathcal{U}_{VPU,pending})$ method. This modularization enables the smooth integration of user defined task scheduling algorithm.

The VPU component for down-stream queuing realizes the delay annotations of the functional tasks as well as additional delay increments in case of task preemption. Again, this functionality corresponds exactly to the $\mathcal{U}_{VPU,projected}$ queue and the related handling functions $f_{VPU,down\_activate}$ and $f_{VPU,down\_update}$.

The local address resolution unit represents a separate block of the VPU, which is not related to the timing model. As outlined in section 6.2.2, this block determines the location of the correct destination address of the transactions initiated by the application tasks. This address resolution is implemented as a hierarchical procedure. First the local address resolution unit checks whether the target task is located on the local VPU. In this case, the transaction is immediately inserted into the up-stream processing unit. Otherwise the global address resolution unit needs to be consulted.

8.3 NoC Framework

This section deals with the unified modeling of the communication architecture to enable a rapid partitioning and dimensioning of the on-chip network. In analogy with the previous section, the NoC framework depicted in figure 8.6 essentially implements the timing manipulation nodes related to communication timing defined in section 7.3.3.

The central module of the NoC simulation framework is represented by the NoC Channel, which captures the generic part of the communication functionality. This comprises e.g. the implementation of the synchronization protocol, the general transaction handling, and keeping track of the user module status.

The communication architecture specific information is encapsulated into a set of network engines. Following the definitions in section 7.3.3, the NoC framework offers three generic engines:

- The **Point-to-Point engine** represents an exclusive resource between a single initiator and a single target.

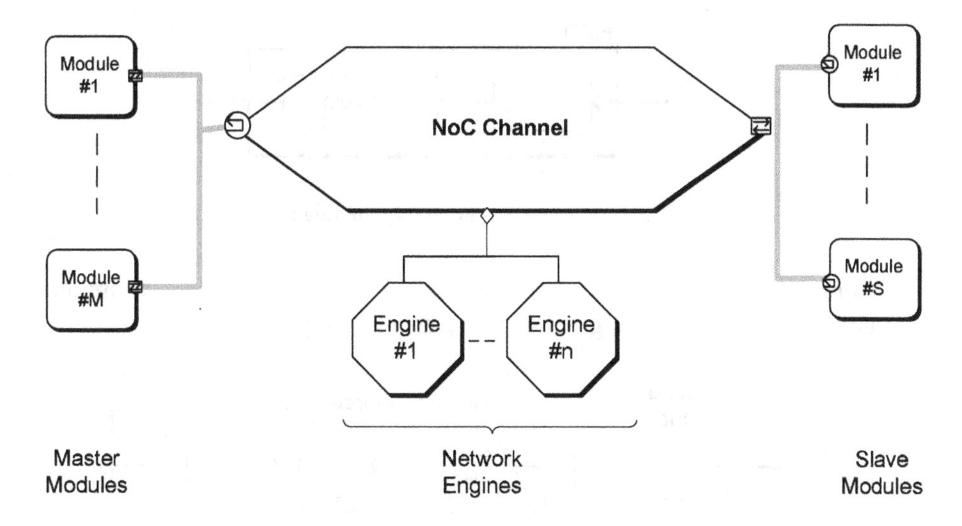

Figure 8.6. NoC Framework Overview

- The **Bus engine** models a shared resource between multiple initiators and targets modules.
- The **Crossbar engine** embodies router like parallel communication resources with a centralized arbitration scheme.

Additionally, a **Hierarchical engine** enables the composition of basic engines to complex network topologies.

8.3.1 Point-to-Point Engine

According to definition 7.16, the point-to-point engine models transactions between source and destination module over exclusive link connections. This engine is configurable by the following performance parameters:

- The **bitwidth** represents the bitwidth of the communication resource.
- The **clock-interval** denotes the clock period of the link.

According to definition 7.27, both parameters together characterize the throughput of the link. Due to the pending queue $\mathcal{U}_{CN,pending}$ in the communication node, the point-to-point engine exhibits the semantics of a FIFO buffer. As shown in figure 8.7, incoming requests are buffered in the FIFO queue as long as the resource is occupied.

8.3.2 Bus Engine

The bus engine models data exchange over a shared bus medium. Similar to the Point-to-Point Engine, the bus engine maintains a queue $\mathcal{U}_{CN,pending}$ to

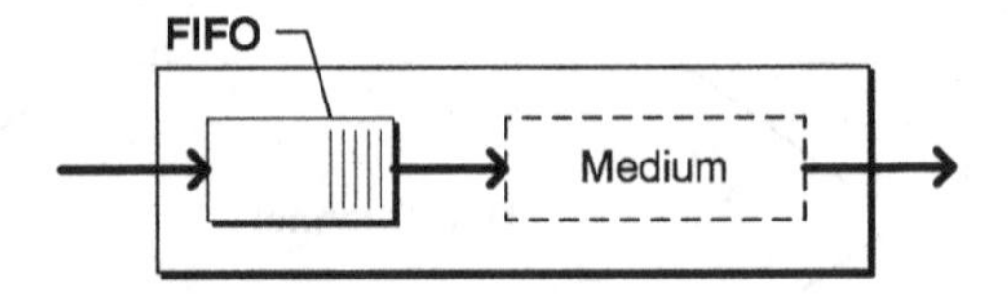

Figure 8.7. Point-to-point Engine

store pending requests. As depicted in figure 8.8, pending requests are queued until the bus resource is granted by the arbitration process.

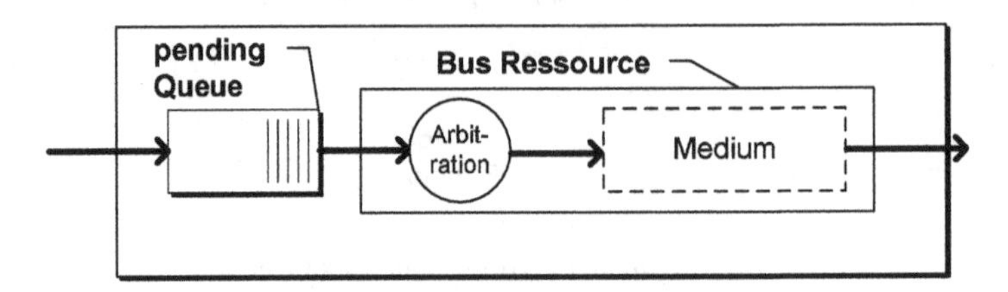

Figure 8.8. Bus Engine

Bus Engines can contain multiple Bus Resources to model the effect of crossbar bus architectures like the STbus [56]. In this case pending requests are dispatched to the appropriate resource. This extension to the bus model with multiple shared resources is visualized in figure 8.9.

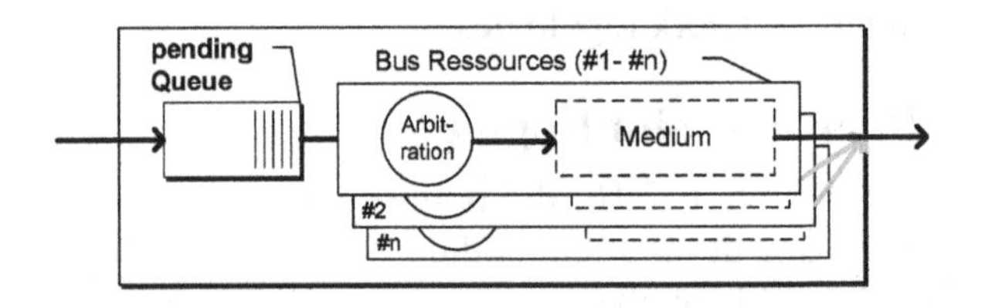

Figure 8.9. Bus Engine with multiple Resources

In order to achieve the accuracy required to take architectural decisions, the bus engine has to be adjustable towards specific bus architectures. As discussed in section 3.3 many different arbitration schemes exist in bus based communication. For this reason the bus engine is designed in a modular way according to the static UML class diagram shown in figure 8.10.

This modularity enables the designer to extend the generic bus engine towards a specific arbitration scheme like for example time division multiple access (TDMA) based bus arbitration. The following units can be extended:

Calculation Unit: This unit calculates the occupation time, that is needed to transfer a token over the exclusive communication resource. To explore basic throughput requirements of the communication architecture, the bus resource is parameterizable with respect to bitwidth and clock-interval:

- The **bitwidth** denotes the bitwidth of the bus resource
- The **clock-interval** denotes the clock period of the bus resource

According to definition 7.17, the base version calculates the transfer time according to the simple formula $\Delta t_{CN,transfer} = \lceil \frac{packetsize}{bitwidth} \rceil * clock - interval.$ Advanced timing calculation units could support preemption of the communication resource, which occurs when burst transactions are intercepted from higher prioritized initiators.

Arbitration Unit: The arbitration unit implements one of the following arbitration algorithms.

- **fixed priority:** Each initiator port has a fixed priority.
- **variable priority:** The priority is determined by the priority field in the packet header.

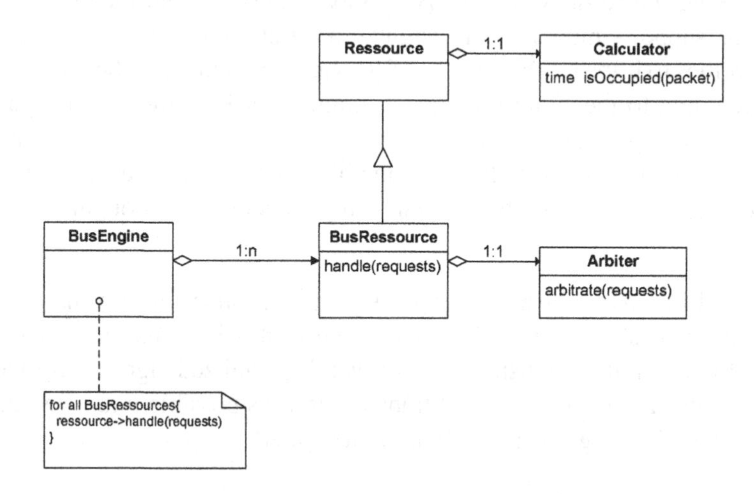

Figure 8.10. Structure of Bus-Engine

- **least recently used (LRU):** The arbiter maintains a history of granted initiators. The RLU arbiter selects the request, which originates from the initiator with the oldest history entry.
- **latency based:** Requests are granted on the basis of preconfigured bandwidth requirements of the initiators.

This unit is configurable by the following set of parameters:

- **arbitration cycles:** number of cycles required to perform the arbitration.
- **priority configuration:** values needed for the selected arbitration algorithm

Thanks to the modular object-oriented structure, the bus engine can be easily enhanced with further arbitration and timing calculation mechanisms by overloading the API of the respective unit. This minimizes the modeling effort for the integration of novel bus architectures into the NoC framework.

8.3.3 Crossbar Engine

The crossbar engine models full-scale on-chip networks, that perform more than one transaction at the same time. Since crossbar networks differ significantly, it is by far not possible to capture all possible incarnations with a single node. The Crossbar Node elaborated below models Virtual Output Queued (VOQ) [74] architectures like e.g. [8] for equal-size data packets with a non-blocking, buffer-less $N_I \times N_O$ crossbar matrix. Additionally, the Crossbar Node supports modeling of weighted arbitration algorithms with P different priorities to improve the Quality of Service capabilities of the on-chip network. According to the VOQ principle, incoming packets are stored separately per output and per priority before they pass through the crossbar matrix. This technique prevents from head-of-line blocking in case of asymmetric traffic or partially blocked outputs and hence improves overall throughput and fairness [3].

- **Arbiter Algorithm** The ingress port controllers inform the arbiter about the states of their VOQs. According to this information, it calculates an I/O-configuration using one of several specialized algorithms to achieve as many simultaneous packet transmissions as possible under certain criteria. The following algorithms are implemented:

[3]Please refer to the introduction of queuing mechanisms on page 28

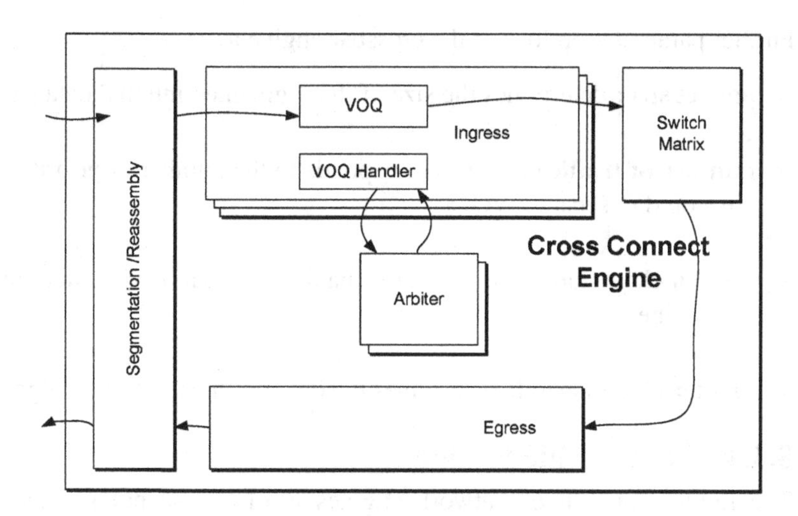

Figure 8.11. Crossbar Engine Overview

- **TDMA Allocation:** static allocation [234], for each time slot a fixed, contention-less matrix configuration is adjusted.
- **PIM:** Parallel Iterative Matching (PIM) is a simple iterative matching algorithm [235], based on random selection of ports.
- **iSlip:** The often employed iSLIP algorithm is a starvation-free algorithm invented by McKeown [236].
- **iLQF:** The iterative longest queue first (iLQF) algorithm has additional capabilities to handle weights.
- **PSLIP:** Parallel version of the iSLIP algorithm [237].
- **SIMP:** Successive Incremental Matching over multiple Ports (SIMP) [238] calculates the matching in one iteration.

- **Weight Generation Algorithm** The status of the VOQ is used by the Arbiter to decide on the arbitration. Different algorithms can be used to calculate the current status value:

  - **Prio:** The weight corresponds to the static priority of the respective VOQ.
  - **OCF:** According to the Oldest Cell First algorithm, the weight depends on the time tag of the head-of-line packet.
  - **LQF:** The Longest Queue First algorithm calculates the weight as function of the VOQ length.

Further parameter options of the crossbar engine are:

- **packet size:** determines the size of the segmented internal data packets,
- **number of traffic classes:** corresponds to the number of priorities, which are stored in separate VOQs,
- **port clock period:** determines the maximum input bandwidth of the crossbar engine,
- **arbiter clock period:** determines the delay of the arbitration algorithm.

8.3.4 Hierarchic Engine

The implementation of network engines is of course not restricted to the engines and algorithms outline above, instead the designer has the full expressiveness of C++ and SystemC to extend the engine library. In particular complex network topologies can be composed from elementary engines by recursively instantiating the NoC Channel inside hierarchical engines. These elementary network engines are linked together by the channel inside the hierarchic engine. Elementary engines are the above mentioned point-to-point engine, the bus engine, the cross-connect engine and all additional user defined implementations.

Internal Composition

An example of a complex meshed network is shown in figure 8.12. The modules Module1 to Module3 are connected by crossbar engines. Module4 and Module5 are peripheral components with lower throughput requirements, which are linked to a local bus.

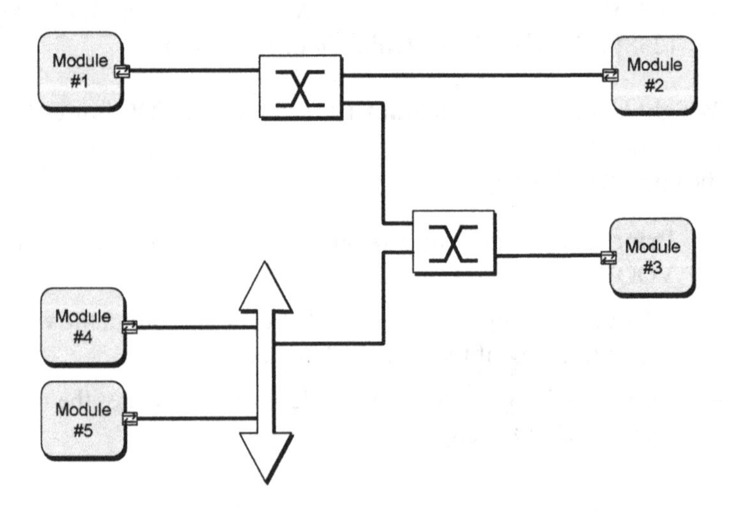

Figure 8.12. Hierarchic Engine Example – Platform View

The following description outlines the NoC framework structure that corresponds to this interconnect example. The additional components implement required functionality, like routing and segmentation and are transparent to the user. As depicted in figure 8.13, this functionality is performed by Network Interface and Network Link components, which are inserted at the network edges and between communication nodes.

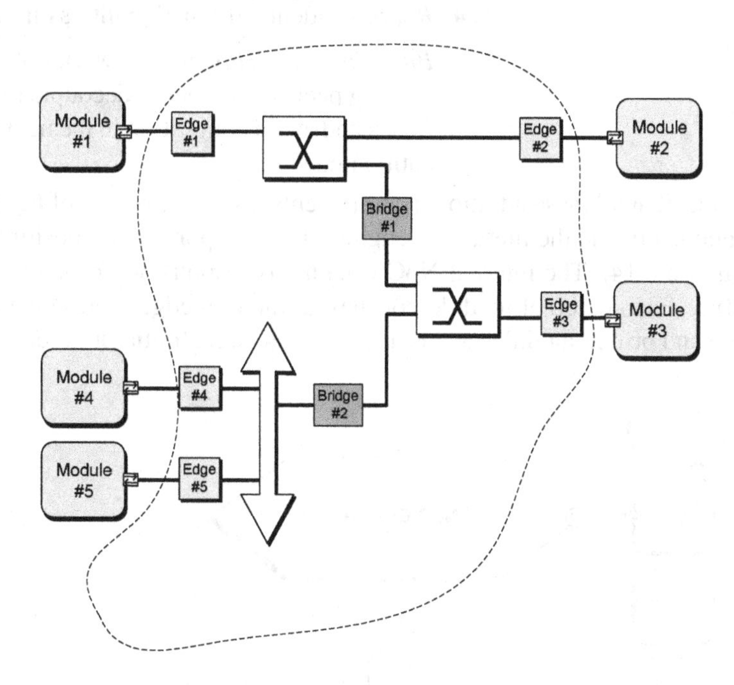

Figure 8.13. Hierarchic Engine Example—Inserted Components

Network Interface (NI) components are added at the edge of the network. Each module access will pass through the NI component assigned to its port. The NI performs two major tasks:

- *Routing:* As further elaborated in the subsequent section, in multi-hop networks the target address needs to be translated to hop specific address ports.
- *Segmentation and Reassembly:* Packets can be segmented in small flits (flow control units) which are sent through the inner network and are automatically reassembled at the destination.

Nework Link (NL) components are added between two inner network nodes. Thus they connect one node to another and are responsible for forwarding packets between network engines. Therefore they can perform the following tasks:

- *Routing:* identical functionality as in the NI.
- *Buffering and delaying:* Network links buffer incoming packets in order to decouple two network nodes and avoid congestion in the next communication node.

Note that all these additional components inside the border of figure 8.13 are contained inside the hierarchic engine. The component composition is shown in figure 8.14. The internal NoC Channel is responsible for network internal traffic. Each external module port has an internal edge representation. These edge and bridge modules act as autonomous target/initiator modules.

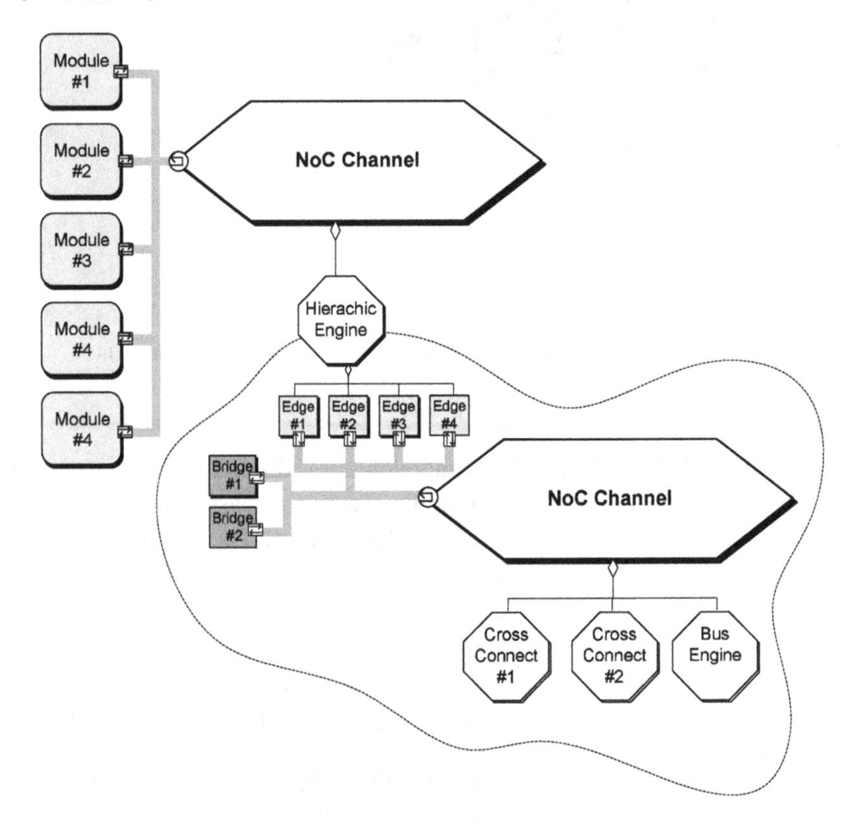

Figure 8.14. Hierarchic Engine Example – NoC Framework View

Source Routing

In general the process of determining and describing the path to forward packets from a source to a destination module is called routing. The routing is fairly simple in topologies with only a single node. When it comes to interconnects with multi-hop topologies each network node has to determine the next network point to which a packet should be forwarded. In network research several procedures like e.g. Label Switching [239] are known which differ significantly in implementation complexity and speed. These issues are not in this scope, because only the pure "path finding" functionality in the simulation model is required. The solution implemented for the NoC framework is called source routing, which simulates efficiently and is easy to configure.

For this purpose each packet contains a list of network node identifiers, which specifies the full path through the network topology. The routing procedure of each network node matches the identifier of the next node and forwards the packet to its respective output port.

Declarative Instantiation

The topology as well as all parameter options of the communication architecture are configured through a set of eXtensible Markup Language (XML) [240] files. Before the actual start of the simulation, these configuration files are parsed and the architectural information is elaborated into an Internal Representation (IR). Based on the information in this IR, all communication nodes and processing elements are instantiated, configured and connected.

In this way, the whole instantiation and binding of the communication architecture is automatically done by the framework. The designer does not need to change the code to evaluate for example a bus instead of a point-to-point link. Only the parameters in the configure file need to be changed. As no recompilation is needed to iterate over different communication alternatives, which dramatically speeds up design space iteration.

8.3.5 NoC Framework Case Studies

The concept of unified communication modeling using the NoC Framework has been applied to real-life bus architectures to evaluate trade-off with respect to modeling efficiency, simulation speed and accuracy. The investigations of the AMBA bus [23, 224] as well as the STBus [241] both demonstrate the fidelity of the outlined approach: A moderate modeling effort of 2-3 weeks is necessary for the creation and verification of a bus specific network engine. Compared to the fully cycle accurate reference model of the respective bus architecture, the

NoC framework achieves an accuracy of 95% in case of the AMBA bus and even 100% for the STBus [4].

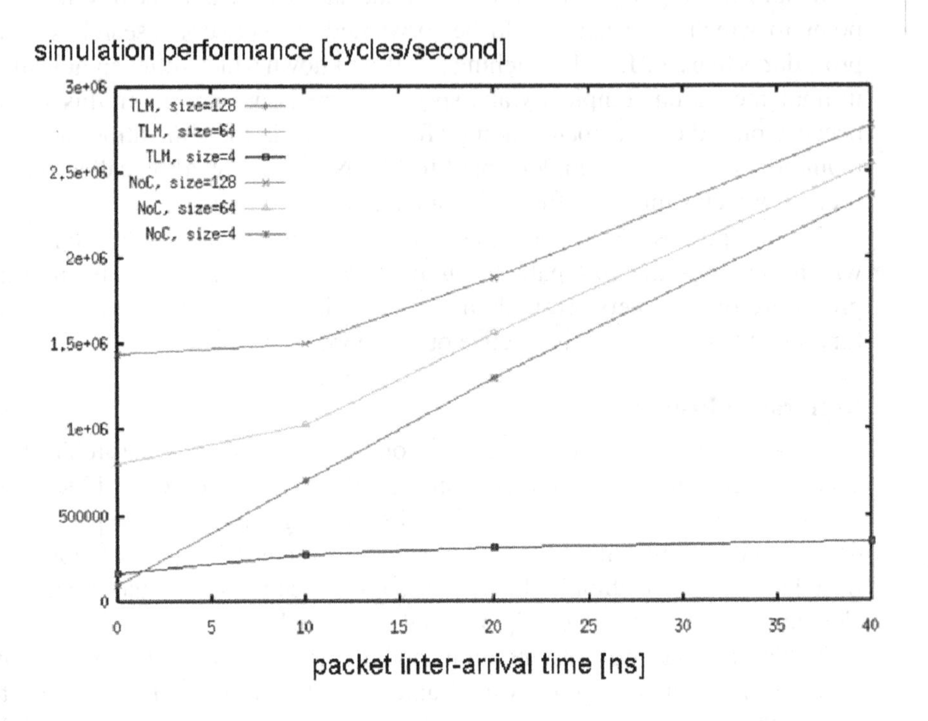

Figure 8.15. Simulation Speed Comparison

Figure 8.15 shows the results of a simulation speed experiment based on a single AMBA bus node with two initiators and two slaves running on a 2.0 GHz linux PC. The simulation speed is plotted over the idle time between successive transactions, which denotes a metric for the utilization of the bus. The different curves highlight the impact of the burst size for the NoC Framework (NoC) as well as for the cycle accurate TLM model (TLM). Naturally the simulation speed of the event-driven NoC Framework is highly dependent on the utilization of the on-chip network as well as the granularity of the communication events, whereas the cycle-driven TLM model is basically not affected by these parameters.

[4] averaged timing derivation for the considered test cases

8.4 Tool Support

This section provides an overview of the visualization tools, which are part of the MP-SoC framework. Due to the immense functional and architectural complexity of MP-SoC platforms, intuitive visualization of the simulation results is of superior importance for the usability of the complete modeling environment.

8.4.1 Trace Views

First the user has to verify the functional correctness of the application model as well as the application to architecture mapping. Since verification of massively parallel applications is not well supported by conventional debugging mechanisms, the MP-SoC framework provides a set of trace views, which visualize the system level activity over time.

MSC Instrumentalization

The MP-SoC framework is equipped with a methodology specific graphical debugger, which visualizes the SystemC simulation according to the Message Sequence Chart (MSC) principle [25].

MSC is a tracing language standardized by the ITU-T [242] for the specification of the communication protocols. An MSC shows the behavior of system components and their environment by means of exchanging messages. This approach is usually employed by design environments based on the Specification and Description Language (SDL) [96], like e.g the TAU SDL suite [243].

As depicted in figure 8.16, the MSC debugger of the MP-SoC framework provides a very intuitive visualization of the events in a coarse-grain SystemC process network. The SystemC modules are displayed by vertical lines, which at the top are labelled with the module name. Communication events are displayed as horizontal arrows between the lines of the initiator and target processes. The arrows are labelled with signal name, the bracketed time instance and the ADT type name. The debugger provides a set of advanced filter mechanisms to systematically reduce the displayed data exchange to the currently interesting communication events [244].

VPU Trace Views

Each VPU instance collects information about the activation of the different tasks in a value change dump (VCD) trace file. Since this file format is commonly used, the dump file can be viewed by different public available tools like the GTK electronic waveform viewer [245]. The screenshot in figure 8.17 below displayes an example of a task activation trace over time.

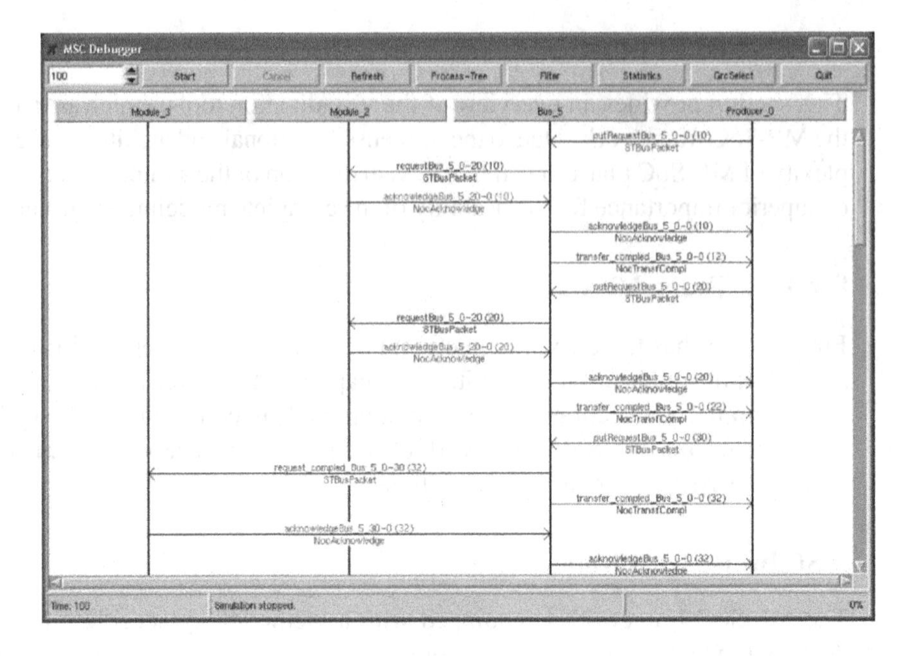

Figure 8.16. Message Sequence Chart Debugger

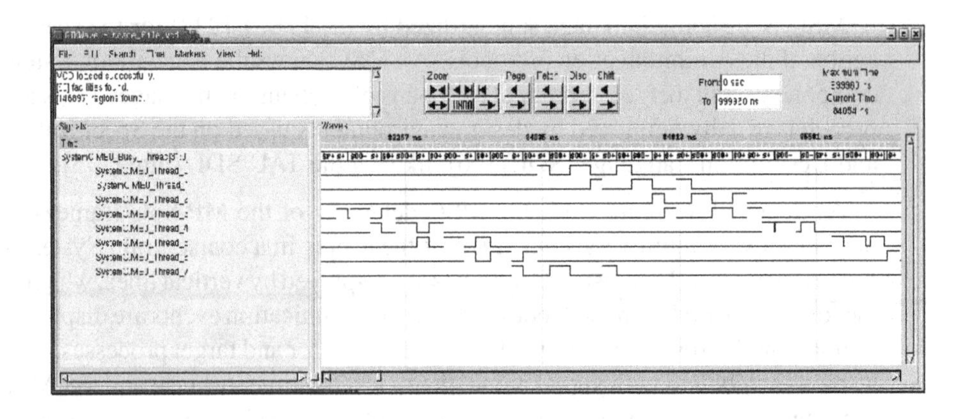

Figure 8.17. VPU Trace View

8.4.2 Statistic Evaluation

After the functional correctness of the system is achieved, the performance of the aniticipated architecture is of major importance to drive the design space exploration process. For this purpose a hierarchy of statistical performance views allows the rapid identification of bottlenecks as well as poor resource utilization.

Section 7.4 defines the various types of performance metrics, which can be derived from the Virtual Architecture Mapping technique. This statistical data is collected automatically through instrumentation of the architecture models, so the focus of this section is on the explanation and interpretation of the generated statistic views. The discussion is separated into the communication and the processing element related performance analysis.

Communication Views

As already formally defined in section 7.4.3, the following metrics allow for a comprehensive analysis of the specified communication architecture:

- The **engine pending time** arises due to resource contention.
- The **engine transfer time** denotes the time needed for the token transfer. Naturally the communication resource is occupied during this period.
- The **target pending time** arises if the target is not ready.
- The **total time** denotes the whole transfer time from the initiator to the target.

In a postprocessing step of the simulation run, these performance metrics are prepared in various levels of detail to enable a hierarchical analysis refinement.

Histogram Views

The histogram is a bar chart representation of the frequency distribution according to definition 7.20, where the height of every bar represents number of observed events. The generated histograms are visualized with the public available gnuplot [246] tool.

A set of example histograms is shown in figure 8.18, where engine pending time, engine duration time, and total time are plotted for two ports.

As depicted in figure 8.19, the NoC framework also aggregates information about module specific communication behavior. These are distributions of packets properties that are sent by the specific module port. These properties are:

- **activation interval:** delta time between two outgoing packets
- **destination Id:** destination port id of the packets
- **packet length:** packet length in bit of the packets
- **request type:** request type 1 is read request; request type 2 is write request

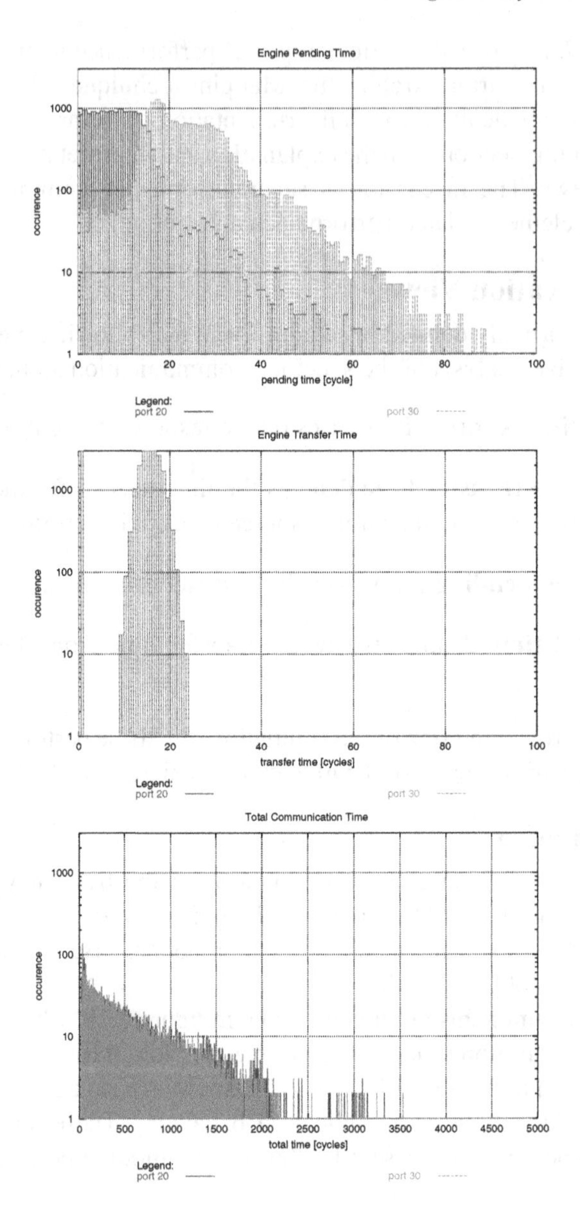

Figure 8.18. Histogram View

Communication Graph Views

The postprocessing procedure also creates a set of graphs that reflect the spatial organization of the specified communication architecture. The graph nodes represent modules and communication resources and the edges denote communication paths. These communication paths can be labeled with different values

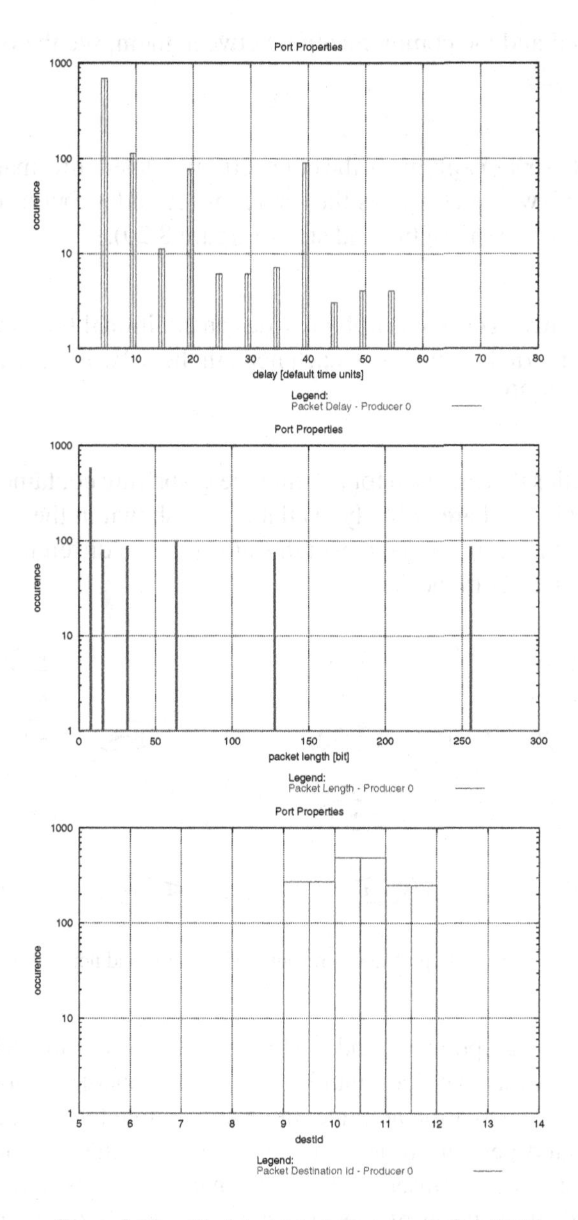

Figure 8.19. Module related Communciation Metrics

like e.g minimum, maximum, average, variance and trials, which are extracted from the histogram distributions. Two different type of graphs are generated:

- The **point-to-point graph** visualizes pure connectivity related information and hides all communciation resources. Only the processing elements are displayed and the communication between them, see the right hand side of figure. 8.20.

- The **network graph** maps the connectivity related information to a network centric view, which shows the interconnect nodes, but hides node-internal resources (see the right hand side of figure 8.20).

- The **resource centric graph** provides an additional level of detail by showing the internal resources of each node in the network (see the left hand side of figure 8.20).

The utilities that are used for plotting the graphs are contained in the graphviz toolsuite [247] and are publicly available. As shown in the example graphs of figure 8.20, the groups of edge weights are coded in different colors to ease the identification of bottlenecks.

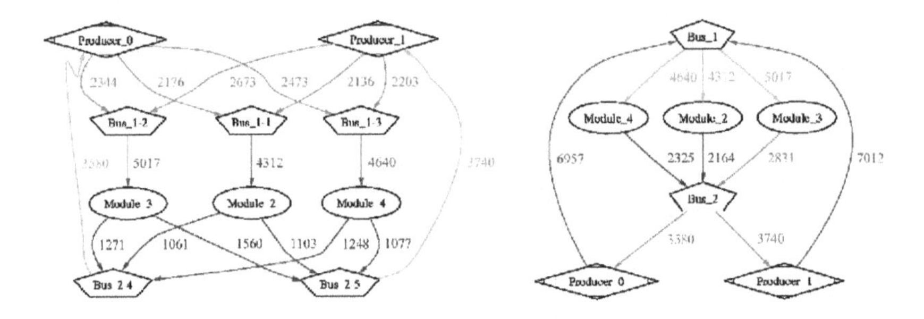

Figure 8.20. Graph Views - resource view (left) and network view (right)

Together the graph views and the histogram views enable a hierarchical performance analysis. The communication graphs enable an assessment of the complete communication architecture. For example the resource centric graph with annotated pending time immediately reveals the communication bottlenecks in the system. Based on this information, the histogram view provides the detailed distribution of the respective parameter on a single connection. This hierarchical analysis refinement is essential to cope with the huge amount of simuation results and by that enable a rapid analysis of the specified communication architecture.

Processing Element Views

Similar to the communication related performance visualization, the MP-SoC framework generates a set of analysis views related to the processing elements.

Delay Annotation Views

As mentioned in section 7.4.1 the analysis of the atomic timing annotation parameters provide insight to the performance of the considered functional process and the required timing budget. Recall, that

- the **processing time aggregation** $\tilde{\Delta}t_{delay}$ of the Initiator Node represents the time that is required by the initiator module to process an incoming event and generate a new request,
- the **initiation interval aggregation** $\tilde{\Delta}t_{init}$ of the Target Node represents the interval unitl a process is ready to accept the next incoming event.

An example histogram is shown in figure 8.21, where processing and iteration interval for two ports are presented.

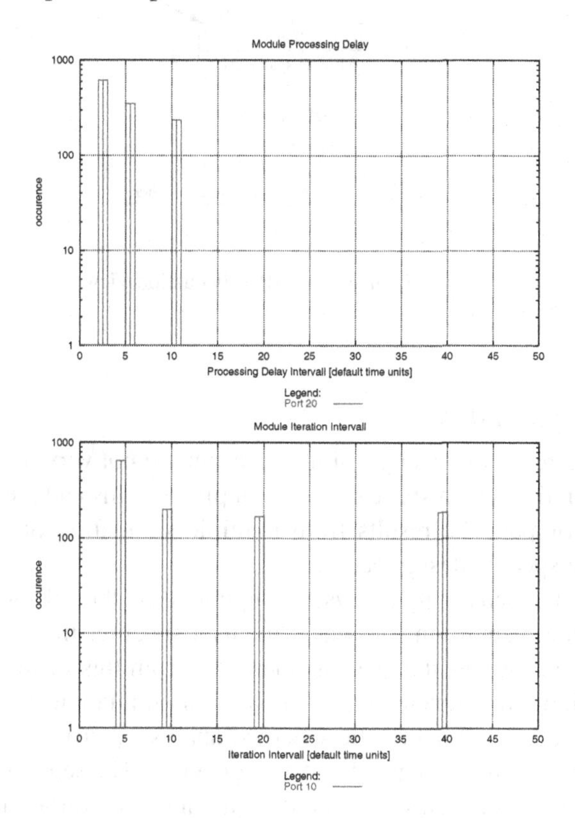

Figure 8.21. processing delay and iteration interval

VPU Evaluation Views

For a detailed analysis of contention in multi-threaded processing elements the number of waiting packets and the waiting time in the VPU pending queue

and VPU projected queue is collected. Thus performance bottlenecks due to unsufficient or excessive instantiation of task.
Again all the collected statistics are aggregated into performance graphs to ease the evaluation of complex distributed systems. Figure 8.22 shows a example plot of the VPU internal communication.

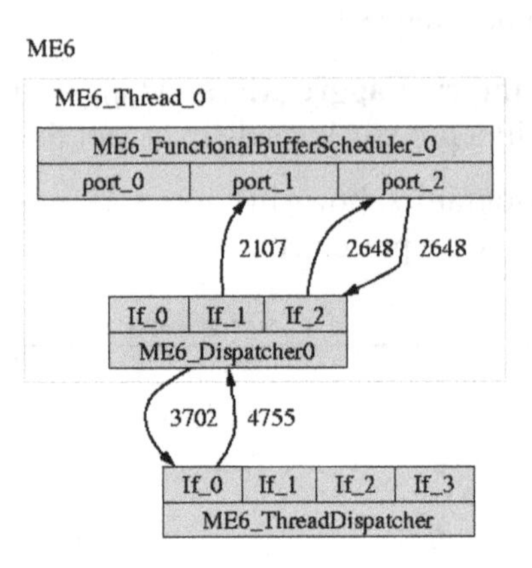

Figure 8.22. VPU Evaluation View

Sensitivity Analysis

Usually the results of a single simulation run are not very meaningful to drive decisions in the design space exploration process. Instead the system architect needs to compare the results from multiple simulation runs to evaluate the impact of a specific design parameter.

This kind of sensitivity analysis is supported by the MP-SoC framework by a configuration environment, where the user specifies the sweeping of design parameters as well as the analysis metrics. From this meta-configuration the configuration environment spawns a batch of simulations to iterate all design parameter settings. Finally, the specified analysis metrics from the individual simulation runs are aggregated into a single view. The sensitivity analysis view plots the analysis metrics over the design parameter variations. The example depicted in figure 8.23 reveals the impact of the design parameter bitwidth of a certain bus node on the application latency.

Sensitivity analysis is an established mathematical discipline and therefore well applicable for further automization [248]. Especially the effort for the initial identification of *significant* design parameters given a set of relevant analysis metrics can be reduced by an automated screening process.

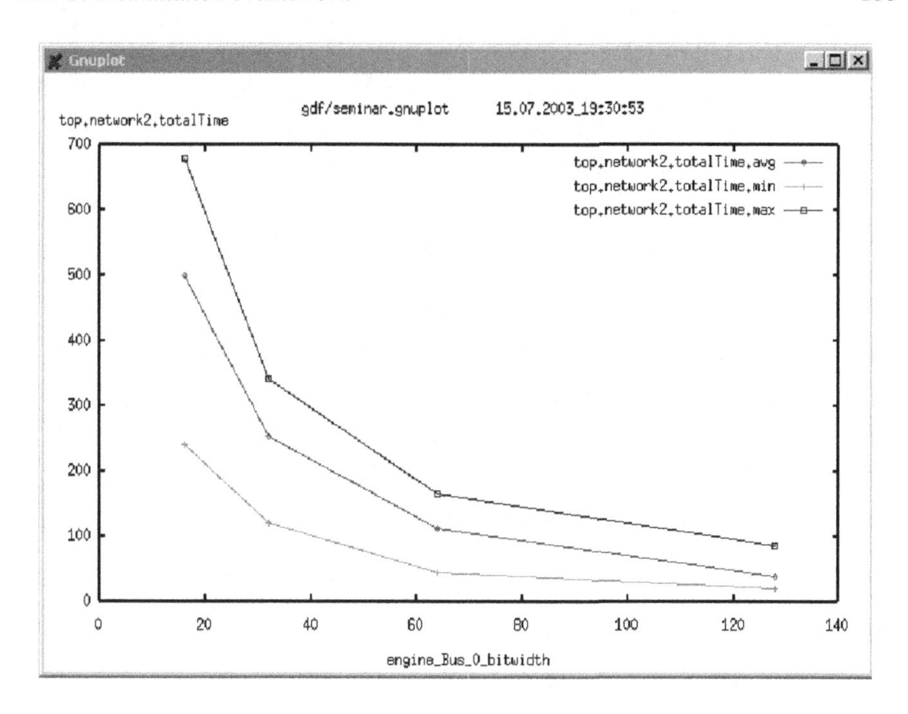

Figure 8.23. Sensitivity Analysis

8.5 Summary

This section describes a modular exploration framework for Network-on-Chip enabled multi-processor platform.

Based on the timing model defined in section 7.3 and the bus agnostic synchronization protocol defined in section 8.1, the major benefit of the outlined approach is the ability to *capture the impact of the anticipated architecture on the system performance in a unified way.* By providing an extendible set of configurable library elements for processing elements and network modules, the MP-SoC framework enables the rapid exploration and coherent comparison of architectural alternatives.

Special emphasis is put on the associated visualization tools, which support intuitive debugging and analysis of most complex MP-SoC platforms.

Chapter 9

CASE STUDY

The acid test for any research on system level design has always been the incorporation into actual design practice. To prevent from incompatibility with real design problems, the MP-SoC framework is by no means a 'one-shot' development. Instead the Virtual Architecture Mapping technology has evolved over multiple industrial design projects in the context of an ATM packet switch [215], a 3D graphics processor [218], and an IP forwarding engine [25] as well as multiple modeling experiments of complex bus architectures [23, 224, 241].

This chapter presents the results from a recent case study, where the MP-SoC framework is employed in the design space exploration and task partitioning of a complex networking application. The selected IPv4 forwarding application is an ideal candidate for the implementation on multi-processor platforms, since it provides abundant data level and task level parallelism. A commercially available network processor unit from Intel serves as a reference platform for the architectural investigations.

After a brief introduction of the selected application and the reference architecture, the discussion is focused on the investigation of architectural alternatives to highlight the potential for optimization by using an efficient design space exploration environment.

9.1 IPv4 Forwarding with QoS Support

The Internet Protocol version 4 is the widely used layer 3 communication protocol for macroscopic computer networks. It represents a typical application from the networking domain as characterized in section 2.1.

The IPv4 DiffServ application depicted in figure 9.1 can be coarsely divided into two major parts: The IPv4 forwarding core application comprises packet parsing, route lookup, and packet fragmentation. The remaining functional blocks are part of the differentiated services mechanism [249], which enhance IPv4 with basic support for Quality of Service (QoS).

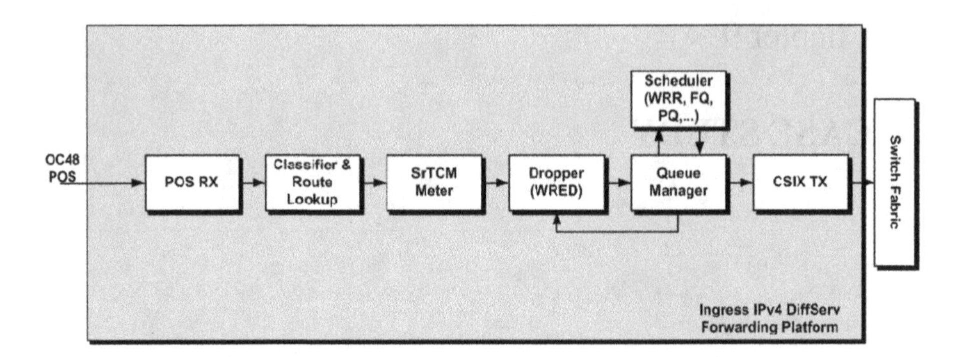

Figure 9.1. IPv4 Differentiated Services Application

The individual functional blocks are briefly introduced in the following paragraphs Delay Annotation

PosRX. This unit receives incoming Packet-Over-Sonet (POS) frames from the physical interface, extracts the IP packet and performs an RFC 1812 [250] compliant 5- or 7-tuple check to validate the header. Afterwords a Packet Descriptor (PD) is generated, which contains the relevant header information from the IP packet. The IP payload is stored in a large DRAM and the all further functional blocks operate on the packet descriptor, which is stored in fast SRAMs.

Route Lookup. The route lookup (RLU) unit determines the next hop address using a longest match table search algorithm.

Classifier. The classifier unit determines the Quality of Service (QoS) class of the actual packet. Together with the RLU this block requires the main computational effort and most of the memory accesses of the DiffServ application.

Meter. The meter verifies that the IP packet flows adhere to the negotiated Service Level Agreement (SLA), which specifies QoS related parameters like guaranteed bandwidth, burstiness, and maximum delay. The implemented Single Rate Meter [251] algorithm marks the packet with green, yellow or red, depending on whether the IP flow satisfies, exceeds or violates the SLA.

Dropper. Depending on the color and current queue fill status, this unit drops packets according to the Weighted Random Early Detection (WRED) algorithm [252] to avoid throughput degradation due to congestion.

Queue-Manager/Buffer. All incoming packets are queued according to their individual traffic class. Dequeuing and forwarding to the CSIX unit is only initiated by the scheduler.

Scheduler. This module decides on the allocation of the available bandwidth to the queued packets. By that the scheduler has a major impact on DiffServ platform in terms of throughput and QoS. Scheduler algorithms differ significantly with respect to fairness, efficiency, worst-case behavior, QoS guarantees, utilization, and computational effort [253]. To enable a coherent comparison of all these parameters during the design space exploration phase, the following parameterizable algorithms are implemented in the scheduler model:

- Fair Queuing (FQ)
- Priority Queuing (PQ)
- Weighted Fair Queuing (WFQ)
- Weighted Round Robin Queuing (WRR)
- Deficit Weighted Round Robin Queuing (WRR)

Please refer to Gries [254] for an in-depth discussion of the individual scheduling algorithms.

CSIX TX. This unit fetches the IP packets from the DRAM memory, segments them into CSIX or SPI-4 compliant frames [255] and transmits the data to the outgoing transmit buffer.

According to the MP-SoC design flow introduced in section 6.3.2, first a functional SystemC model of the IPv4 DiffServ application is created. The application tasks are represented as reactive processes and all inter-task communication is performed via the generic synchronization interface.

The validation of completeness and functional correctness is performed by means of a reactive process network according to definition 7.11. This representation does not yet impose any assumptions on the architectural realization, but serves as the starting point for the subsequent Virtual Architecture Mapping.

9.2 Intel IXP2400 Reference NPU

The IXP2400 network processor is a member of Intel's second-generation network processor family [256] and represents a true Multi-Processor SoC platform. The IXP2400 employs eight fully programmable and hardware multi-threaded processing elements connected by an heterogeneous communication network.

IXP2400 Overview

The block diagram of the IXP2400 network processor architecture depicted in figure 9.2 shows the following units:

- The **Intel XScale** micro controller unit performs control-plane processing and resource management.
- The **eight Micro Engines (MEs)** are specialized for data-plane processing of networking tasks. Each micro-engine features eight concurrent hardware threads to hide memory access latencies.
- The **SRAM Controller** provides two independent QDRII[1] compliant channels with a peak bandwidth of 12.8 Gbit/sec each to access the on-chip scratch memory as well as the off-chip SRAM
- The **DRAM Controller** provides channel with a peak bandwidth of 19.2 Gbit/sec to the off-chip DRAM
- The **heterogeneous On-Chip Network** comprises three global buses to hook the micro-engines and the XScale core to the memory resources. Additionally adjacent micro-engines are connected by next-neighbor registers for rapid passing of data and state informations.

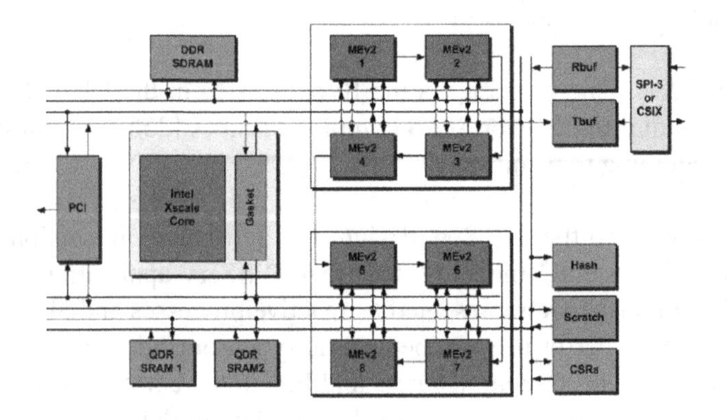

Figure 9.2. Intel IXP2400 Network Processor Block Diagram

The IXP2400 provides sufficient performance and flexibility to execute a wide variety of high performance applications such as multi-service switches,

[1]QDR: Quad Data Rate

DSLAMs (DSL access multiplexers), CMTS (cable modem termination system) equipment, 2.5G and 3G wireless infrastructure and Layer 4-7 switches including content-based load balancers, and firewalls.

IPv4 Mapping to IXP2400

The goal of the subsequent design space exploration experiments is to evaluate custom MP-SoC platforms for the IPv4 DiffServ application against the IXP2400 reference architecture. To achieve a fair comparison between architectural alternatives, the performance characterization of the functional IPv4 forwarding tasks are taken from the original Intel documentation [14]. The individual timing annotations and communication requirements are listed in table 9.1.

Table 9.1. Characterization of the IPv4 DiffServ Application Tasks.

	compute cycles (Δt_{init})	SRAM accesses	DRAM accesses	total cycles
Parser	70	3	2	945
RLU	160	10	1	1660
Meter	80	2	0	330
Dropper	80	2	0	330
Buffer	> 10	4	0	500
Scheduler	100	0	0	100
Segmentation/CSIX	80	3	1	705

In the same way, the mapping of the tasks to the processing elements of the IXP2400 depicted in figure 9.3 is aligned to the Intel reference implementation [14]. The initial allocation of IXP2400 Hardware threads to the application tasks results from matching the individual computation and communication characteristics of the individual tasks to the available resources under consideration of the top-level application requirements:

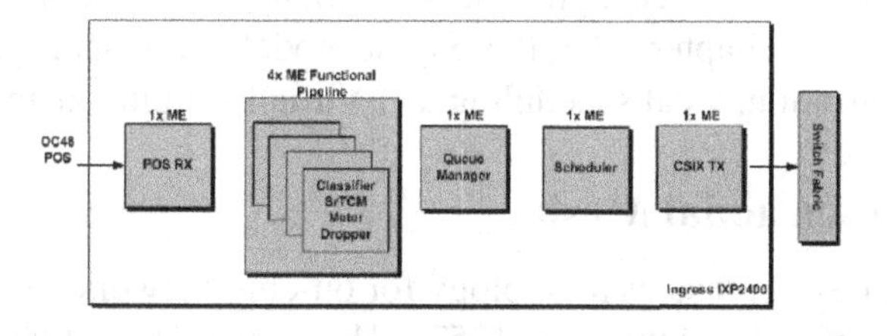

Figure 9.3. Reference Mapping of IPv4 DiffServ Application to IXP2400 Platform

Today's IPv4 router typically support an OC-48 data rate, which corresponds to 2.5 Gbit per second. The resulting inter-arrival time between two back-to-back minimum sized POS packets (40 byte + 6 byte Point-to-Point(PPP) protocol overhead) is 147ns. All the processing and communication has to be performed within this budget, which corresponds to 88 cycles on the 600MHz micro-engine. Naturally this can only by achieved by exploiting the task level parallelism and hiding the memory access latency using Hardware Multi-Threading [14].

Using the unified timing modeling defined in chapter 7, the functional application model has been mapped to a virtual architecture model of the IXP2400 architecture. This comprises the annotation of the processing requirements as listed in table 9.1 to the reactive process models of the application tasks. Additionally, the NoC framework is used to instantiate the bus and point-to-point communication resources of the IXP2400 on-chip network. The generic VPU is used to simulate the behavior of the eight micro-engines, where each VPU emulates eight concurrent Hardware threads.

Before the simulation results are presented, the following section discusses architectural alternatives for the on-chip communication network.

9.3 Custom IPv4 Platform

This section discusses alternative communication architectures to replace the shared bus architecture of the IXP2400 platform. The motivation behind these experiments is that future MP-SoCs will employ on-chip networks to address physical and functional shortcomings of the shared bus paradigm. Investigated network architectures are a global crossbar resource, a 2-dimensional Mesh (2D-MESH), an Octagon-Grid and an application specific micro-network, which exploits the locality of the IPv4 DiffServ application.

Global Crossbar

The first step to overcome bandwidth limitations is to replace the global bus resources with a central crossbar architecture depicted in figure 9.4.

Using the NoC framework, the architecture model is easily adjusted to the new architecture by exchanging the bus resources with one cross connect engine. As described in chapter 8.3.3, this engine models a full-scale packet switch with global arbitration, local scheduling and Virtual Output Queuing (VOQ).

Two-Dimensional Mesh

The 2D-Mesh represents a topology for on-chip networks, which is often proposed in NoC related literature [257]. This structure is a two dimensional grid in shape of a chessboard, where adjacent network nodes are connected via network links. Figure 9.5 shows the micro-engines, memories and I/O components of the IXP2400 platform hooked to the Mesh communication architecture.

Using the NoC Framework, this kind of complex topologies are realized using the hierarchic engine described in section 8.3.4. In case of regular topologies like the 2 dimensional mesh and the octagon, the creation of the NoC Framework configuration files for the instantiation of the communication resources can be largely automated using topology generators [223, 129].

Octagon Network

The Octagon structure represents an on-chip backbone network based on a hypercube topology, which specifically addresses the high throughput requirements of network processors [78]. Compared to the 2D-Mesh grid, the increased connectivity between the distributed nodes reduces the average communication latency, since each node can reach every other node with a maximum of two hops. Figure 9.6 shows the IXP modules connected to the Octagon network structure.

Local Crossbar

In contrast to the general on-chip networks in the previous sections, the local crossbar topology is tailored to the traffic requirements of the IPv4 forwarding application.

A closer investigation of the memory access profile of the functional tasks listed in table 9.1 reveals, that only the PosRX and CSIX blocks are accessing the DRAM memory. On the other hand side, the RLU/CLASSIFIER, Meter, Dropper and Buffer blocks are accessing heavily the SRAM/SCRATCH memories, but never the DRAM. Therefore the processing elements executing the PosRX and CSIX functionality should be close to the DRAM. Accordingly, RLU/CLASSIFIER, Meter, Dropper, and the Buffer should be close to the

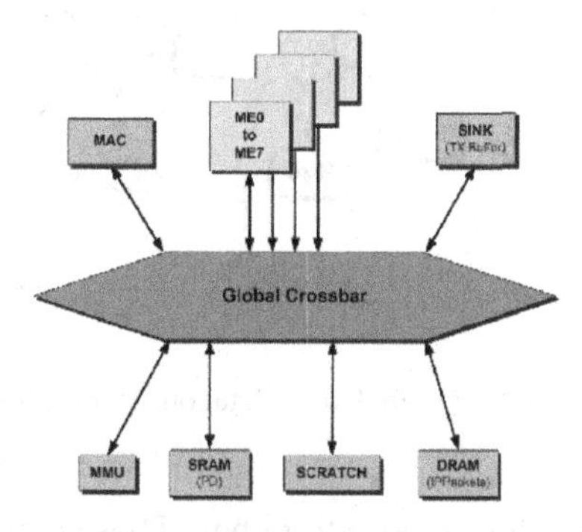

Figure 9.4. Global Crossbar Architecture

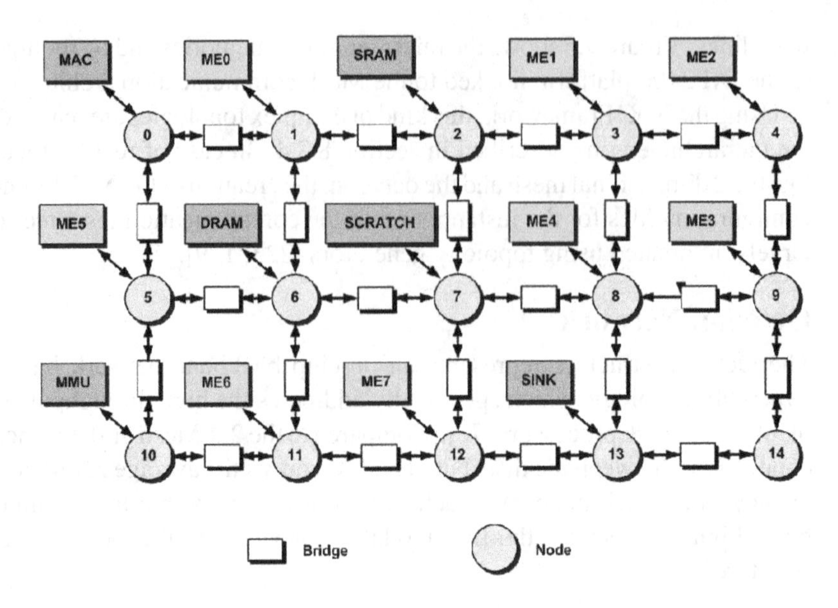

Figure 9.5. Two-Dimensional Mesh Architecture

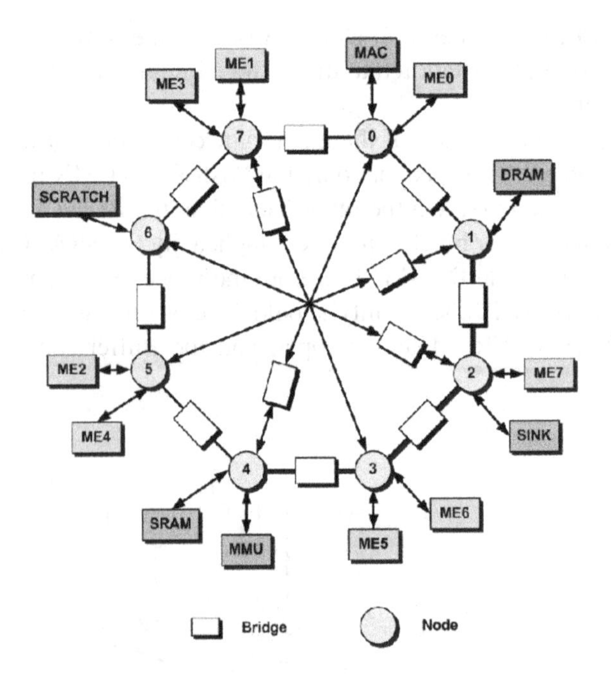

Figure 9.6. Octagon Architecture

SRAM/SCRATCH blocks close together. These considerations are reflected in the application specific network architecture depicted in figure 9.7.

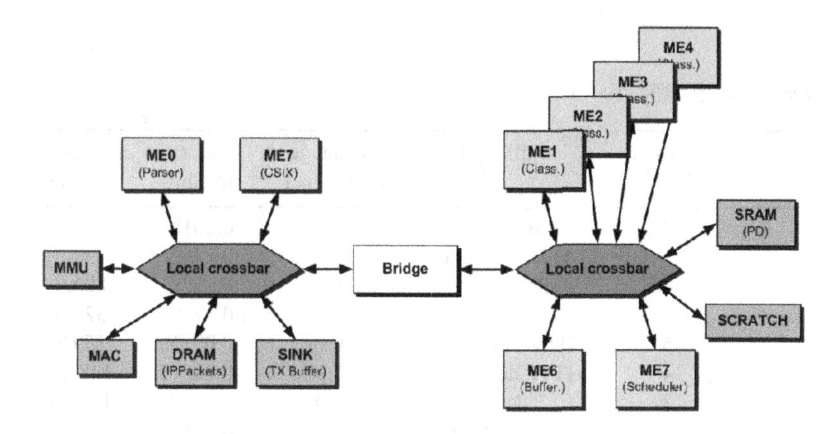

Figure 9.7. Local Crossbar Architecture

9.4 Simulation Results

The goal of the following design space exploration experiments is to reveal architectural trade-offs and highlight the potential for optimization. Special emphasis is put on the correlation of application and architecture related aspects as well as the correlation between the communication network and the processing elements. A detailed discussion of the architectural alternatives is beyond the scope of this book [2].

Simulation Scenario

The stimulation of the architecture model is of crucial importance to obtain meaningful simulation results. In this case the input stimuli for the IP forwarding processor are generated by a set of statistical traffic sources [258]. In analogy with [254], these traffic generators are parameterized to create a typical IP traffic scenario.

Table 9.2 shows the different traffic classes, the content type, the proportional contribution to the total traffic, the data rate, and the average size of the IP packets. The decreasing traffic classes reflect the decreasing Quality of Service requirements of the different packet flows.

Performance Impact of the Communication Network

A highly condensed extract of the simulation results is presented in table 9.3. The first two columns of the table define the given system with its communication architecture and the number of concurrent Hardware threads per micro-engine. The next column contains the number of parallel communication

[2] please refer to [223] for an in-depth presentation and analysis of the simulation results

Table 9.2. Traffic classes.

traffic class	traffic type	percentage of traffic load	data rate in Gbps	packet size in bytes
9	CBR voice	9.3	0.2018	128.0
8	Video P-frame	11.2	0.2452	238.2
7	Video I-frame	5.2	0.1128	896.6
6	Signaling	8.1	0.1758	320.0
5	HTTP Request	4.2	0.0911	159.8
4	HTTP download	37.1	0.8051	1480.1
3	FTP download	10.0	0.2170	1497.5
2	Transaction	5.6	0.1215	276.2
1	Flooding	9.3	0.2018	40.6

resources in the communication network. The last column denotes the available bandwidth per communication resource, which is required to meet the application level performance requirements. For example in case of 8 Hardware threads, the Octagon network requires a bandwidth of 12.8 Gbps to sustain the OC-48 IP traffic throughput.

Table 9.3. Simulation Results.

line	communication architecture	threads per processing element	parallel resources	required bandwidth in Gbit/s
0	ixp2400	8	2 / 1 / 7	12.8 / 19.2 / 32
1	crossbar	8	8	12.8
2	octagon	8	16	12.8
3	custom	8	10	12.8
4	crossbar	16	8	3.2
5	Octagon	16	16	6.4
6	custom	16	10	3.2

Line 0 of table 9.3 reflects the simulation results of the Intel IXP2400 reference architecture, consisting out of 8 micro-engines each implementing 8 threads for computation. As presented in section 9.2, the communication architecture comprises two bus resources connected to the SRAM, one bus resource for the DRAM, and 7 next-neighbor connections. Stimulated with the IP traffic scenario from table 9.2 this system achieves a throughput of approximately 2.14 Gbit/s. Together with the resulting IP latency of approximately 14000ns, the measured performance of the IXP2400 platform defines the reference point for the comparison with the following platform configurations.

Line 1 of table 9.3 presents the global crossbar system. After replacing all communication resources of the reference architecture with one global crossbar switch, the system requires a bandwidth of 12.8Gbit/s per switchable resource to meet the application performance of the reference platform.

Lines 2 and 3 of table 9.3 summarize the measurements obtained from the distributed communication topologies. In total the Octagon based network contains 16 parallel communication resources, which require a bandwidth of 12.8Gbit/s to meet the reference performance. As illustrated in figure 9.7, the custom network contains two crossbars connected via a bridge. This configuration requires the same bandwidth per resource as the octagon and global crossbar platforms to meet the performance of the reference architecture, however the number of parallel resources is below the octagon topology.

Performance Impact of Hardware Multi-Threading

So far only the effect of different communication architectures has been evaluated. Using the generic VPU, also the number of concurrent threads per processing element can be easily modified. The second part of table 9.3 is dedicated to the combined evaluation of computation and communication architectures. The results in line 4 to 6 present the above introduced systems with an additional change of the VPU configuration. The threads of each micro-engine have been doubled from 8 to 16 threads each to explore the impact of changes in both parts of the designs.

Duplicating the number of threads per micro-engine in the global crossbar system (line 4) improves the systems performance significantly. The required bandwidth to achieve the performance of the reference architectures can be decreased to 3.2Gbit/s, which is equal to a 75% decrease in required bandwidth.

Similarly the octagon system (line 5) improves by duplicating the number of threads per micro-engine. Contrary to the crossbar system the improvement turns out to be slightly less significant, due to the higher latency on the more distributed communication architecture.

Finally the custom system in line 6 achieves a high performance together with a distributed architecture. The required bandwidth per parallel resource is 3.2Gbit/s to achieve a performance equal to the reference architecture from Intel.

These results demonstrate the advantage of a framework for the joint investigation of the processing elements and the communication architecutre.

Increasing the threads per processing element greatly relieves the constraints on the interconnect architecture. These type of trade-offs can have a significant impact on the quality of results in terms of chip size, energy efficiency and performance.

Chapter 10

SUMMARY

For the foreseeable future, embedded application domains like wireless communications, multimedia and networking will continue to push the development of integrated circuits to it's economic and technological limits.

Economically, the development of state-of-the art integrated circuits is a significant endeavor. The mask fabrication cost accounts for about one million US dollar and the development cost can be more than one order of magnitude higher. Additionally, the short life-cycles of today's embedded products imposes a huge pressure on the time-to-market and time-in-market to secure the return of investment.

From the technical point of view, the algorithmic complexity of next generation embedded applications even out-paces the advances in semiconductor technology. Apart from these demanding performance requirements, rapidly evolving and differentiating standards call for a new level of flexibility to proliferate the time-in-market of integrated circuits. On top of this, the increase in performance and flexibility has to be delivered without impacting the energy consumption, as battery capacity remains fairly constant.

Looking at the architectural implementation for these kind of demanding embedded applications, the functional complexity clearly promotes Software enabled solutions to achieve the required flexibility and to cope with the demanding time-to-market conditions. However, the stringent power dissipation constraints of mobile applications and cost sensitive consumer devices prohibit the use of general purpose processors. Instead, the tight cost and performance requirements of versatile embedded systems lead to application specific heterogeneous Multi-Processor SoC architectures [259].

Heterogeneity applies to all architectural elements of the MP-SoC platform. The processing elements have to be individually tailored to the respective application task to meet performance, flexibility, and energy efficiency constraints [11]. Similar considerations lead to an application specific partitioning of the communication and memory organization into a *clustered* platform architecture. Here each processing element accesses its local cache and scratch-pad memories via short high-speed bus architectures. On the other hand, global memory access and data exchange is handled by dedicated on-chip networks [7].

Problem Statement

The outlined evolution of embedded applications and architectures will soon lead to a design complexity crisis, as state-of-the-art EDA tooling and methodologies are not shaped to address the development of heterogeneous MP-SoC platforms. Today there is a gap in the design flow between the architecture agnostic algorithmic development and the block-level implementation domain.

The only available kind of System-Level Design tools addressing this gap are based on cycle-accurate Transaction-Level Modeling, which is still way too detailed to cope with the architectural design complexity of next generation MP-SoC platforms. The major obstacle for efficient design-space exploration on this level are the target architecture specific interfaces of the communication and processing models. This requires a significant modeling effort to change for example the communication from a shared bus to an on-chip network. Similarly, every time the processing element changes the application has to be ported to the new target architecture. Apart from the modeling effort the limited simulation speed hinders the large-scale exploration of design parameter options.

Contribution

In the course of this book, a design methodology and corresponding tool environment has been developed, which enables the joint consideration of application and architectures above cycle accuracy. The higher level of abstraction allows the efficient definition of complex clustered MP-SoC architectures as well as the spatial and temporal mapping of the application.

The entry point for the developed design methodology is a coarse-grain communicating process network representing the application. This representation preserves the application inherent task level parallelism and exhibits the inter process communication. The communication is based on an architecture independent synchronization interface based on the Open Core Protocol (OCP) semantics.

The key contribution of this book is the *Virtual Architecture Mapping* mechanism, which enables the transparent mapping of the application model to the anticipated MP-SoC architecture. The idea behind this mechanism is to map the un-timed application space events onto timed architecture space events. These architecture space events reflect the execution of the application tasks on the processing and communication resources available in the anticipated MP-SoC platform.

The event mapping problem has been separated and solved in three categories.

- The OCP synchronization protocol has been extended with a concise set of primitives for *timing annotation*, which reflect the individual processing resource requirements of a single application task. These concepts have recently been incorporated into the new OCP TL3 standard API [150] (see appendix C).

- A generic *Virtual Processing Unit* models the concurrent execution of multiple application tasks on a shared processing element. The VPU captures the impact on performance of both Software Operating Systems as well as Hardware Multi-Threading. This enables the rapid exploration of spatial and temporal application mappings to multi-processor platforms.

- The *Network-on-Chip framework* provides a generic mechanism to integrate arbitrarily complex communication architecture models. Based on three configurable communication nodes for exclusive point-to-point resources, shared buses and complex cross-connects, the on-chip communication can be easily changed from a global bus scheme to a clustered network with complex topologies and multiple hierarchy levels. The NoC framework is now commercially available as an option of the CoWare Platform Architect (see appendix C).

To identify the system bottlenecks and to perform trade-off decisions, the system architect needs to be equipped with meaningful evaluation data. The developed framework automatically creates a database with aggregated simulation statistics. To cope with the huge amount of analysis information, two hierarchical level of statistical views are generated: A set of communication graphs provides a system level view of a specific analysis metric like e.g. throughput, latency or contention. For detailed analysis the generated resource level histograms reveal in more detail the distribution of the respective metric. For verification issues the framework supports system-level tracing by means of interactive Message Sequence Chart (MSC) representations as well as Value Change Dump (VCD) files.

Case Studies

The effectiveness of the developed design methodology, modeling framework and evaluation tools has been demonstrated by a number of complex design project in the context of networking and graphic processing applications.

In particular the modeling of several real-world communication architecture like AMBA AHB, STbus and Philips AEthreal has validated the high accuracy achievable with abstract Transaction Level Modeling.

The capabilities of the complete framework have been proven in a case study, where the Intel IXP2400 networking platform is employed in the context of an IP forwarding application with Quality of Service support. This complex MP-SoC platform comprises 8 multi-threaded processors, each implementing 8 concurrent threads, as well as further peripheral modules.

During this case study several communication architectures and VPU configurations have been defined, simulated and quantitatively compared against the IXP2400 reference architecture. This experiments have validated the successive refinement methodology for the optimization of the communication architectures. Additionally, the results reveal the system-level trade-offs between the communication architecture and multi-threaded processing elements.

Recently the technology has been transferred to CoWare Inc, where the concepts described in this book have been productized and incorporated into the CoWare Platform Architect product line (see appendix C).

Outlook

Given the evolving complexity of embedded applications and MP-SoC architectures, the concepts developed in the course of this book and the implemented MP-SoC modeling framework can only serve as starting point for further investigations. The following sections list a number of opportunities for future research and development activities.

Tool Related Improvements

System efficiency metrics are always a relation on performance and cost related functions like power and area. Future versions of the MP-SoC framework could also take an estimation of power and area into account. These values can be elaborated on basis of the known topology and of the data that is collected during the simulation.

Further Applications and Case Studies

Further case-studies like e.g. modeling of the Sonics MicroNetwork, the Artheris NoC, or the new ARM AXI bus architecture, would increase the confidence into the new methodology and extend the library of IP models.

A very useful application of the developed NoC framework would be the creation of a comprehensive 'NoCStone' benchmark suite. As a first step, the communication characteristics for a set of target applications have to be extracted as an input specification for workload generation. Together with a library on on-chip communication architectures, the NoC framework can generate a matrix of benchmark results, which reveals the applicability of a communication architecture for a certain application domain. However, significant expert knowledge is required to consider the right parameter settings for both the workload generators as well as the architecture models.

Extended Software Flow

SystemC 2.1 enables a Software centric modeling style by supporting dynamic task creation and completion. Currently the MP-SoC framework is clearly focused on an architecture exploration use model. In principle the concept of Virtual Architecture Mapping can be integrated into an embedded Software development flow to enable early consideration of task level parallelism and HW/SW partitioning.

For this purpose, the VPU has to be extended with RTOS modeling capabilities like task control as well as Software specific communication and synchronization primitives. Apart from this, a library of Software related parameterizable components need to be established containing, e.g. cache, memory management unit, interrupt controller, and direct memory access.

Verification Aspects

Besides the systematic refinement flow, the value of the proposed MP-SoC design space exploration framework also reaches into the verification of subsequent implementation steps. The information contained in the architecture specification files serves as a *predictor for the verification flow* [260] to verify the functional correctness of error-prone implementation models as well as to review the estimated timing properties.

In that the approach takes the synchronous design principle, which enabled the RTL based modeling, to the next level: transistor-level timing is neglected during the specification at the register transfer level. After the physical design, the clock synchronization points enable the verification of the assumed timing budgets and physical gate- and wire-delays.

In close analogy, the packet-level events serve as synchronization points, which can be recovered at the cycle-accurate implementation level. By that, the timing annotations in the MPSoC framework can be automatically translated into assertions.

Appendix A
The OSCI TLM Standard

In March 2005 the OSCI Transaction Level Modeling (TLM) working group has released version 1.0 of the TLM Kit. This chapter gives a brief overview about the essential aspects of the TLM standard. Please refer to [142, 233] for a more detailed introduction.

The Transaction Level Modeling standard is built as a set of interfaces that define the basic semantics for the communication between models. These basic TLM primitives provide with the fundamental communication and synchronization constructs that can be used to create TLM models. Although these interfaces can be used in their primitive form, it is generally expected that they will need to be complemented with a *protocol-layer*. As depicted in figure A.1 this intermediate layer will target the basic interfaces to a specific on-chip communication protocol or design task. The protocol layer can provide with a set of convenience functions that create a much easier to use TLM API. The Programmers View and Architects View API are exemplary representatives of this protocol layer.

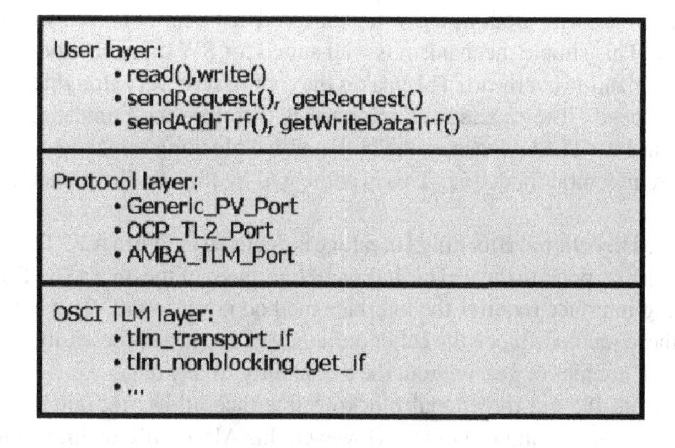

Figure A.1. TLM Communication Layers

The value of the TLM 1.0 standard is in the definition of the basic semantics and terminology. This in principle fosters interoperability between TLM models, because user-level TLM protocols built on-top of the TLM standard can be mapped to a well-defined low-level interface. This eases the creation of transactors between different user-protocols. The TLM working group is currently working on improving the interoperability by driving the standardization up to the user protocol layer.

The basic TLM API consist of bidirectional and unidirectional interfaces each with different possible synchronization mechanisms. Note that in the context of SystemC the terms *blocking* and *non-blocking* have a particular meaning. A blocking interface implies that this interface has to be called from within an SC_THREAD, as such the implementation of the interface is allowed to contain wait(.) statements. In contrast a non-blocking interface cannot contain a wait(.) statement since it is allowed to call such an interface from within an SC_METHOD which is not capable of performing the context switch that is required to implement the wait(.) call.

The Bi-directional Interface depicted in figure A.2 provides with the most simple communication and synchronization mechanism. It uses a REQ data type for the information an initiator provides in the communication and a RSP data type for the information it receives from the target. The synchronization is such that the initiator expects the response packet to be available when the interface call returns. Therefore interface communication never fails and all communicated information is available with every call.

```
template<typename REQ, typename RSP>
class tlm_transport_if : public virtual sc_interface {
public:
  virtual RSP transport(const REQ&) = 0;
};
```

Figure A.2. Bidirectional Blocking TLM Interface

The `transport` interface corresponds to the simple Remote-Procedure-Call (RPC) paradigm: The caller invokes a method, which is implemented in the callee and only returns after the method is executed. This simple mechanism is well suited for SW oriented modeling, because it avoids any synchronization overhead. This makes the coding style very straight forward and yields high simulation speed. The `transport` interface is therefore the foundation for the Programmers View use case for TLM. However, the bidirectional blocking semantics limit the capabilities for accurate architectural modeling. This is addressed by the uni-directional interfaces.

The Uni-Directional Blocking Interface is depicted in figure A.3. The `put` and `get` methods essentially correspond to the `write` and `read` interfaces of the `sc_fifo`. The synchronization of the blocking interface requires the interface method never to fail. That is to say, the implementation of these methods block the caller until they can return successfully. A reason for blocking would be the attempt to `get` without the availability of any data.

In principle the uni-directional blocking interface allows the modeling of protocols with pipelining of requests and responses. However, this API is still too limited to model cycle accurate TLM protocols.

```
template < typename T >
class tlm_blocking_get_if : public virtual sc_interface {
public:
  virtual T get( tlm_tag<T> *t = 0 ) = 0;
  virtual void get( T &t ) { t = get(); }
};

template < typename T >
class tlm_blocking_put_if : public virtual sc_interface {
public:
  virtual void put( const T &t ) = 0;
};
```

Figure A.3. Unidirectional blocking TLM interface

The non-blocking interfaces depicted in figure A.4 are slightly more complex. The `nb_put` and `nb_get` methods execute always immediately and therefore return a boolean value to indicate success or failure. The additional `nb_can_put` and `nb_can_get` methods indicate that data to get or space to put is available. To avoid constant polling, the non-blocking interfaces also provide an event which is notified whenever data or space becomes newly availability.

```
template < typename T >
class tlm_nonblocking_get_if : public virtual sc_interface {
public:
  virtual bool nb_get( T &t ) = 0;
  virtual const sc_event &ok_to_get( tlm_tag<T> *t = 0 ) const = 0;
  virtual bool nb_can_get( tlm_tag<T> *t = 0 ) const = 0;
};

template < typename T >
class tlm_nonblocking_put_if : public virtual sc_interface {
public:
  virtual bool nb_put( const T &t ) = 0;
  virtual const sc_event &ok_to_put( tlm_tag<T> *t = 0 ) const = 0;
  virtual bool nb_can_put( tlm_tag<T> *t = 0 ) const = 0;
};
```

Figure A.4. Unidirectional non-blocking TLM interface

This API is usually sufficiently expressive to model arbitrary user-level TLM protocols [233, 150].

Appendix B
The OCPIP TL3 Channel

The concepts of the generic synchronization interface outlined in sections 6.1.3 and 8.1 have been incorporated into the newly defined OCP TL3 API. The complete documentation of this API can be found in [150]. This appendix gives an update on the changes between the generic synchronization interface and the OCP TL3 API.

The OCP TL3 Interface

The goal of the TL3 API is to enable unified modeling of arbitrary applications and architecture platforms at an abstraction level above cycle accuracy. In that the scope of the TL3 API is clearly beyond modeling of systems, which deploy OCP as the actual bus protocol.

In essence the TL3 API combines two aspects:

First, the set of communication primitives represents the *generic subset of the OCP protocol*. This boils down to a simple dual-way handshaking, where the producer initiates a transaction, which then has to be accepted by the consumer. A transaction, which is sent out but not yet accepted is considered to be *in progress*.

Second, the API provides support for *timing annotation* to promote the dual-parameter timing model[1]. The additional advantage is that the usual code for modeling timing is factored into the channel implementation.

The master interface is separated into a request and a response part, which are depicted in figures B.1 and B.2 respectively. The master request interface represents the *producer* side of the request channel, whereas the master response interface represents the *consumer* side of the response channel.

[1] Please refer to section 6.2.1 for an introduction and section 7.3.1 for a formal definition

```
template <typename REQ>
class OCP_TL3_MasterRequestIF : virtual public sc_interface
{
public:
  virtual bool sendRequest(const REQ& req) = 0;
  virtual bool sendRequest(const REQ& req, const sc_time& time) = 0;
  virtual bool sendRequest(const REQ& req, const int cycles) = 0;
  virtual bool sendRequestBlocking(const REQ& req) = 0;

  virtual bool requestInProgress(void) const = 0;

  virtual const sc_event& RequestStartEvent(void) const = 0;
  virtual const sc_event& RequestEndEvent(void) const = 0;
};
```

Figure B.1. OCP TL3 Master Request Interface

The producer interface in figure B.1 comprises one basic and three derived methods for initiating a transaction. Two of the derived methods provide timing annotation to delay the actual sending of the request for a certain amount of time. The third derived method provides a blocking interface, which is more convenient in case SC_THREADs are used. Whether or not the request channel is currently available can be checked using the `requestInProgress` method.

Additionally the producer interface provides two events, which are notifi ed at the start and end of a request transaction respectively.

```
template <typename RESP>
class OCP_TL3_MasterResponseIF : virtual public sc_interface
{
public:
  virtual bool getResponse(RESP& resp) = 0;
  virtual bool getResponseBlocking(RESP& resp) = 0;

  virtual bool acceptResponse(void) = 0;
  virtual bool acceptResponse(const sc_time& time) = 0;
  virtual bool acceptResponse(const int cycles) = 0;

  virtual bool responseInProgress(void) const = 0;

  virtual const sc_event& ResponseStartEvent(void) const = 0;
  virtual const sc_event& ResponseEndEvent(void) const = 0;
};
```

Figure B.2. OCP TL3 Master Response Interface

The consumer interface in figure B.2 comprises a blocking and a non-blocking method to retrieve the current request. The accept method to signal the completion of the current transaction is available in un-timed and timed variants. The two events and the `responseInProgress` method have the same semantics as in the producer interface.

Note that the request and response channels are completely symmetric. It is therefore sufficient to discuss the master side of the API definition. The slave request interface represents the consumer side of the request channel and is completely analogous to the master response interface. In the same way the slave response interface resembles the master request interface.

The OCP TL3 Channel

The OCP TL3 implementation is realized as a user-level protocol layer on top of the OSCI TLM API to demonstrate the compliance with the TLM standard (see appendix A). As depicted in figure B.3, all the functions and events in the OCP TL3 API are mapped to the non-blocking unidirectional OSCI TLM standard.

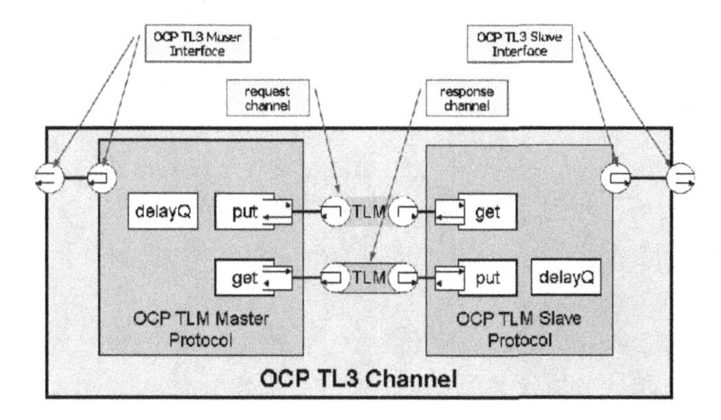

Figure B.3. Mapping the TL3 API onto the OSCI TLM standard

From the outside perspective, the TLM based OCP channel looks just like any other OCP channel, i.e. it implements a master and a slave interface. Inside the OCP TLM channel the interfaces are implemented by two separate modules, the master-protocol and the slave-protocol.

The master- and slave- protocol entities are completely separated, i.e. they communicate only via two OSCI TLM FIFOs. In that the mapping of the OCP TL3 API onto the TLM API is 100% complete. Each of the TLM FIFOs is of size 1, which corresponds to the current transaction in the OCP channel.

The timing annotation features of the TL3 API require additional functionality. The timed sendRequest (and sendResponse) methods are implemented using a delay queue as specified in definition 7.5 on page 86. These queues delay the sending of transactions by the specified amount of time. The timed `requestAccept` (and `responseAccept`) methods are simply implemented by means of delayed event notification.

The OCP TL2 Channel

The OCP TL2 [illegible] API to demonstrate the compliance of [illegible] Figure B.[illegible] the functions [illegible] the OCP TL2 [illegible] on the [illegible] original OSCI TL2 standard.

[illegible]

[illegible]

[illegible]

Appendix C
The Architects View Framework

In 2004, the MP-SoC simulation framework technology has been transferred from the ISS institute to CoWare Inc. From then on the technology has been productized and incorporated into the CoWare Platform Architect product line. This appendix gives a brief overview of the resulting Architects View Framework (AVF) and highlights the relation between the product features and the concepts described in this book.

Overview

AVF is now a commercially modeling environment for architectural exploration of highly complex Multi-Processor SoC platforms. It supports the standardized OCP TLM communication interfaces at the TL2 and TL3 abstraction levels. These protocol agnostic APIs enable a unified representation of communication architectures, which in turn leads to a highly flexible architecture exploration process. The TL2 and TL3 APIs provide support for timing annotation according to the timing model introduced section 6.2.1 and formally defined in section 7.3.1.

The AVF modeling methodology is consistent with CoWare's overall TLM strategy. Transaction-level models created for other use-case like SW development and verification can be reused for architectural exploration. This is achieved by means of bus-transactors, which serve as a performance overlay model to refine the timing information of behavioral platform models to the level required for architectural exploration [153, 150].

Debugging

The value of the Message Sequence Chart (MSC) view for system-level debugging of complex TLM models has been demonstrated by the graphical debugger prototype described in section 8.4.1. As depicted in figure C.1, MSC debugging is now incorporated into the Eclipse based Integrated Development environment.

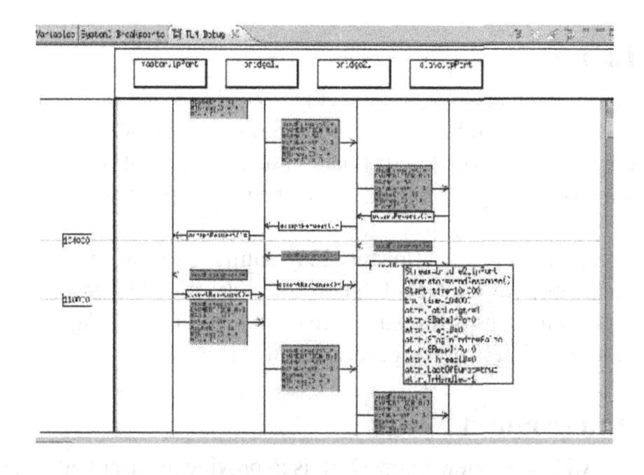

Figure C.1. Message Sequence Chart based TLM Debugging in the CoWare SystemC IDE

Dynamic Configuration

One of the key features in the Architects View Framework (AVF) is the support for declarative platform composition (please refer to page 129).

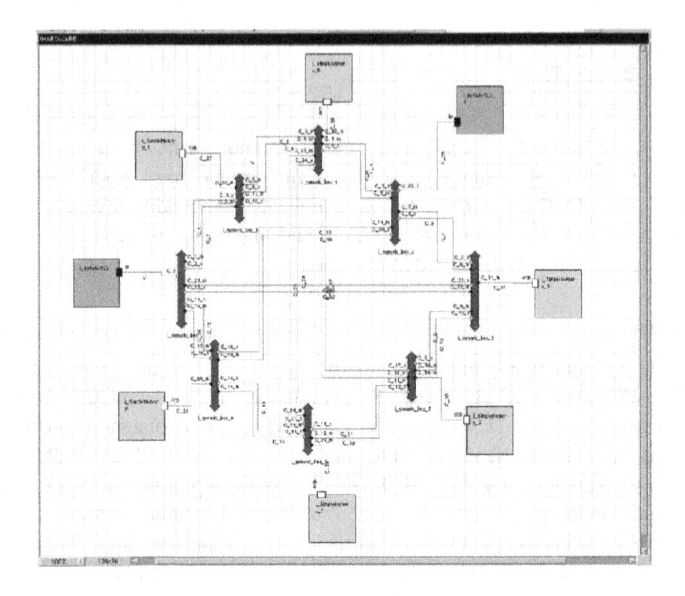

Figure C.2. Platform Assembly and Configuration in CoWare Platform Creator

As depicted in figure C.2, and AVF platform can be assembled, connected, and configured like a regular platform in the graphical Platform Creator tool. Usually the platform is then exported as a static SystemC netlist. The new *AV export* feature exports the platform as a *dynamic netlist*. The Platform Creator tool generates a generic simulation model and a xml confi guration file containing the platform specific information. For any change in the platform architecture only the xml file needs to be re-exported. This cuts the lengthy re-compilation of the simulation model from the exploration cycle.

AVF Bus Library

Apart from the tool enhancements, the actual models of the interconnect architecture are an important aspect of the Architects View Framework. The AVF Bus Library is a collection of highly configurable generic communication nodes for point-to-point, bus, and router based communication. The configurability refers to functional aspects as well as to performance related aspects.

The configurability of the functionality like routing, queuing, and arbitration is similar to the configurability in nodes of the NoC Framework as described in section 8.3. The timing model of the communication nodes is formally defined in section 7.3.3. Figure C.3 depicts a subset of the configuration parameters of the generic bus node.

Modeling Interconnect Nodes

One goal of the Architects View Framework is to provide the user with a modeling environment for the creation of interconnect models. For this purpose the AVF bus library is organized in several layers as depicted in figure C.4.

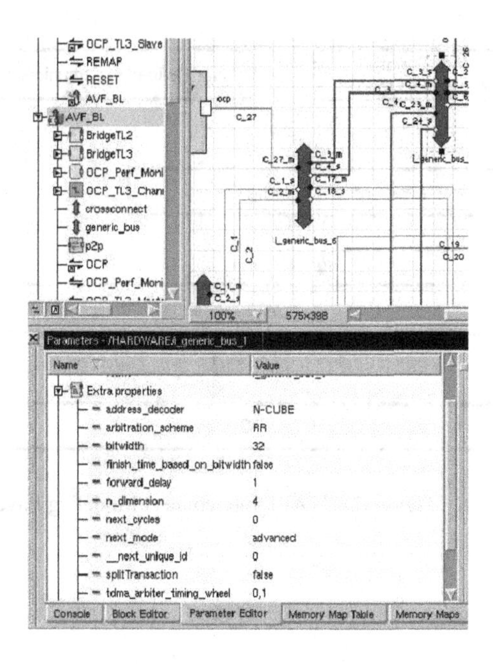

Figure C.3. Configuration Parameters of the AVF Generic Bus

The highest layer is represented by the ready-to-use AVF bus library of configurable interconnect models. By tuning the configuration parameters the generic interconnect models can represent a particular interconnect architecture (see section 8.3.5).

In case the structure of the interconnect architecture differs significantly from the generic models in the AVF bus library, the user can deploy the AVF Interconnect Modeling Objects to create a more customized model with reasonable effort. These modeling objects resemble the components of the communication timing model as defined in section 7.3.3, like e.g. objects to model queuing, arbitration, and timed communication resources.

In case the Interconnect Modeling Objects are still not flexible enough to model the interconnect architecture to the desired degree of accuracy, the user can fall back to the AVF-Scheduler API. This level corresponds to the basic elements for modeling timed process networks as defined in the first paragraph of section 7.2. In particular the AVF-Scheduler API provides an efficient delay queue mechanism to ease the modeling of timing and pipelining.

In any case, the interconnect models composed inside the AVF Modeling Framework are automatically instrumented with analysis and debug hooks (e.g. Message Sequence Chart). Additionally, the underlying AVF Scheduler minimizes the interaction with the SystemC scheduler to optimize the simulation speed.

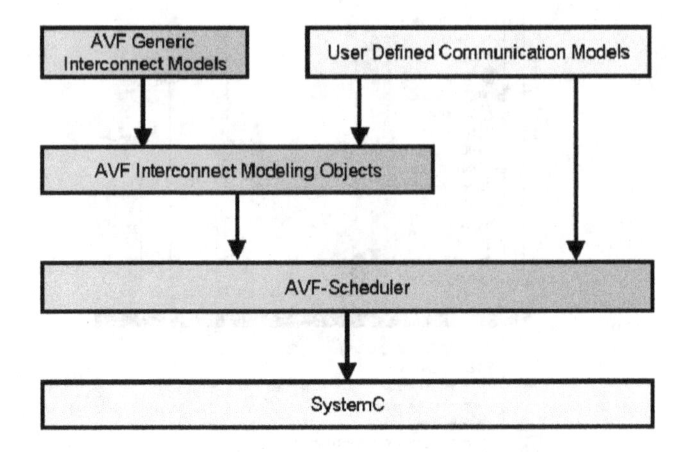

Figure C.4. AVF Interconnect Modeling Layers

List of Figures

List of Tables

References

[1] R. Georing. Innovation Drives SoC Performance, Jul 2005. www.eetimes.com.

[2] Semiconductor Industry Association. *International Technology Roadmap for Semiconductors (ITRS)*, 2001. http://public.itrs.net.

[3] H. Chang, L. Cooke, M. Hunt, G. Martin, A. McNelly, L. Todd. *Surviving the SOC Revolution - A Guide to Platform-Based Design.* Kluwer Academic Publishers, 1999.

[4] T. Kogel, H. Meyr. Heterogeneous MP-SoC – The Solution to Energy-Efficient Signal Processing. In *Proc. of the Design Automation Conference (DAC)*, June 2004.

[5] G. Martin. *System-on-Chip and Network-on-Chip Design*, chapter 16. CRC Press, 2006. ISBN 0-8493-2824-1.

[6] W. Wolf. A Decade of Hardware/Software Codesign. *IEEE Computer*, pages 38–43, April 2003.

[7] L. Benini, G. De Micheli. Networks on Chips: A New SoC Paradigm. *IEEE Computer*, pages 70–78, January 2002.

[8] K. Goossens, J. van Meerbergen, A. Peters, P. Wielage. Networks on Silicon: Combining Best-Effort and Guaranteed Services. In *Proc. Int. Conf. on Design, Automation and Test in Europe (DATE)*, 2002.

[9] Arteris Unveils Strategy, Technology for enabling Network on Chip (NoC) Design. Press Release, March 2003.

[10] K. Goossens. Systems on Chip, Personal Computers or Correct Performance. presented at the MPSOC conference, July 2005.

[11] T. Gloekler, H. Meyr. *Design of Energy-Efficient Application-Specific Instruction Set Processors.* Kluwer Academic Publishers, 2004. ISBN 1-4020-7730-0.

[12] G.P. Fettweis. Embedded vector signal processor design. In *Proc. Int. Workshop on Systems, Architecturs, Modeling and Simulation(SAMOS)*, July 2003.

[13] T. Ungerer, B. Robic, J. Silc. A Survey of Processors with Explicit Multithreading. *ACM Computing Surveys*, 35(1):29–63, March 2003.

[14] S. Lakshmanamurthy, K.-Y. Liu, Y. Pun, L. Huston, U. Naik. Network Processor Performance Analysis Methodology. *Intel Technology Journal*, 6(3), August 2002.

[15] M. J. Rutten, J. T. J. van Eijndhoven, E. G. T. Jaspers, P. van der Wolf, O.P. Gangwal, A. Timmer, E.-J.D. Pol. A Heterogeneous Multiprocessor Architecture for Flexible Media Processing. *IEEE Design & Test of Computers*, 19(5):39–50, July-August 2002.

[16] C.J. Glossner, T. Raja, E. Hokenek, M. Moudgill. A Multithreaded Processor Architecture for SDR. *Proceedings of the Korean Institute of Communication Sciences*, 19(11):70–85, November 2002.

[17] T. Agerwala. Systems Trends and their Impact on Future Microprocessor Design. Keynote of 35th Annual International Symposium on Microarchitecture, Nov 2002.

[18] T. Grötker, S. Liao, G. Martin, S. Swan. *System Design with SystemC*. Kluwer Academic Publishers, 2002.

[19] Platform Architect. *CoWare, http://www.coware.com.*

[20] CoCentric System Studio. *Synopsys, http://www.synopsys.com.*

[21] E. Verhulst. *Beyond the von Neumann Machine: Communication as the Driving Design Paradigm for MP-SoC from Software to Hardware*, chapter 11, pages 217–238. Kluwer Academic Publishers, 2003.

[22] Open Core Protocol International Partnership (OCP-IP). *OCP datasheet, http://www.ocpip.org.*

[23] M. Ariyamparambath, D. Bussaglia, B. Reinkemeier, T. Kogel, T. Kempf. A Highly Efficient Modeling Style for Heterogeneous Bus Architectures. In *Proc.IEEE Int. Symp. on System-on-Chip (SoC)*, November 2003.

[24] T. Kogel, M. Doerper, A. Wieferink, R. Leupers, G. Ascheid, H. Meyr, and S. Goossens. A Modular Simulation Framework for Architectural Exploration of On-Chip Interconnection Networks. In *CODES+ISSS*, October 2003.

[25] T. Kogel, A. Wieferink, R. Leupers, Gerd Ascheid, H. Meyr, D. Bussaglia, M. Ariyamparambath. Virtual Architecture Mapping: A SystemC based Methodology for Architectural Exploration of System-on-Chip Designs. In *Proc. Int. Workshop on Systems, Architecturs, Modeling and Simulation(SAMOS)*, July 2003.

[26] E.A. Lee, A. Sangiovanni-Vincentelli. A Framework for Comparing Models of Computation. *IEEE Transactions on Computer-Aided Desig of Integrated Circuits and Systems*, 17(12):1217–1229, Dec 1998.

[27] H. Zimmermann. OSI Reference Model - The ISO Model of Architecture for Open Systems Interconnection. *COM*, 28(4), April 1980.

[28] P.G. Paulin, C. Pilkington, E. Bensoudane. StepNP: A System-Level Exploration Platform for Network Processors. *IEEE Design & Test of Computers*, 19(6):17–26, Nov-Dec 2002.

[29] K. Goossens, O.P. Gangwal, Jens Röver, A.P. Niranjanz. Interconnect and memory organization in SOCs for advanced set-top boxes and TV- evolution, analysis, and

trends. In J. Nurmi, H. Tenhunen, J. Isoaho, A. Jantsch, editor, *Interconnect-Centric Design for Advanced SoC and NoC*, chapter 15, pages 399–423. Kluwer Academic Publishers, 2004.

[30] T.A.C.M. Claasen. High Speed: Not the Only Way to Exploit the Intrinsic Computational Power of Silicon. In *In Proceedings of the International Solid-State Circuits Conference*, 1999.

[31] Grady Booch, Ivar Jacobson, and James Rumbaugh. The unified modeling language (uml) for object–oriented development. UML Document Set 1.1, Rational Software Corporation, September 1997.

[32] B. Stroustrup. *The C++ Programming Language*. Addison Wesley, third edition, 1997.

[33] K. Arnold, J. Gosling, D. Holmes. *The Java Programming Language*. Addison Wesley, third edition, 2000.

[34] K. Keutzer, S. Malik, A.R. Newton, J.M. Rabaey, A. Sangiovanni-Vincentelli. System-level design: Orthogonalization of concerns and platform-based design. *IEEE Transactions on Computer-Aided Desig of Integrated Circuits and Systems*, 19(12):1523–1543, December 2000.

[35] M. Grammatikakis, M. Coppola, F. Sensini. *Software for Multiprocessor Networks on Chip*, chapter 14, pages 281–303. Kluwer Academic Publishers, 2003.

[36] J.M. Paul. Programmers' Views of SoCs. In *CODES+ISSS*, October 2003.

[37] Michael Flynn. Very high-speed computing systems. *Proceedings of the IEEE*, 54:1901–1909, December 1966.

[38] D.M. Tullsen, S. Eggers, H.M. Levy. Simultaneous Multithreading: Maximizing On-Chip Parallelism. In *Proceedings of the 22th Annual International Symposium on Computer Architecture*, 1995.

[39] M. Forsell. A scalable high-performance computing solution for networks on chips. *IEEE Micro*, 22(5):46–55, Sep-Oct 2002.

[40] J. Goodacre. ARM Multiprocessing. Int. Symp. on System-on-Chip (SoC), November 2003. Invited Talk.

[41] D. Seal. *ARM Architecture Reference Manual*. Pearson Education, second edition, December 2000.

[42] MIPS Inc. Multithreading boosts processor efficiency. Press Release, 2003.

[43] J. Glossner et al. Future Wireless Convergence Platforms. In *Proc. of the IEEE/ACM/IFIP Int. Conference on Hardware/Software Codesign and System Synthesis*, September 2005.

[44] M. Adiletta, M. Rosenbluth, D. Bernstein, G. Wolrich, H. Wilkinson. The Next Generation of Intel IXP Network Processors. *Intel Technology Journal*, 6(3), August 2002.

[45] Agere Systems Inc. *http://www.agere.com*.

[46] Applied Micro Circuits Corporation. *http://www.amcc.com*.

[47] IEEE. *VME Bus standard is IEEE Std 1014-1987.*

[48] PCI-SIG. *PCI Local Bus Specification Rev. 2.2.* http://www.pcisig.com/specifications/conventional/conventional_pci.

[49] J.L. Ayala, M. Lopez-Vallejo, D. Bertozzi, L. Benini. *State-of-the-Art SoC Communication Architectures*, chapter 20. CRC Press, 2006. ISBN 0-8493-2824-1.

[50] D.B. Gustavson. *J. Zalewski (Ed.), Advanced Multiprocessor Bus Architectures*, chapter Computer Buses – A Tutorial, pages 10 – 25. IEEECSP, 1995.

[51] S.J.A. Zaidi, M. Ou, L.E. Adams, H.I. Ramlaoui, B.D. Mills, R. Bhagat. Chip-core framework for systems-on-a-chip, July 2003. United States Patent No. 6,601,126.

[52] VSI Alliace. *VSI Alliance - Architecture Document*, version 1.0 edition.

[53] D. Flynn. AMBA: enabling reusable on-chip designs. *IEEE Micro*, 17(4):20–27, July-Aug 1997.

[54] W. Remaklus. On-chip bus structure for custom core logic designs. In *WESCON/98*, pages 7–14, Sep 1998.

[55] R.A. BErgamaschi, W.R. Lee. Designing Systems-on-Chip Using Cores. In *Proc. of the Design Automation Conference (DAC)*, 2000.

[56] STMicroelectronics, http://www.stmcu.com/inchtml-pages-STBus_intro.html. *STBus Communication System - Concepts and Definitions*, May 2003.

[57] G. Pelissier, R. Hersemeule, G. Cambon, L. Torres and M. Robert. Bus analysis and performance evaluation on a SOC platform at the system level design. Technical report, Universite Montpellier II, STMicroelectronics, 2001.

[58] Drew Wingard. Micronetwork-based integration for SOCs. In *Design Automation Conference*, pages 673–677, 2001.

[59] K. Lahiri, A. Raghunathan, G. Lakshminarayana. LOTTERYBUS: A new high-performance communication architecture for system-on-chip designs. In *Proc. of the Design Automation Conference (DAC)*, 2001.

[60] Vesa Lahtinen, Kimmo Kuuslinna, Tero Kangas, Timo Hämäläinen. Interconnection Scheme for Continuous-Media Systems-on-a-Chip. *Microprocessors and Microsystems*, 26:123–138, 2002.

[61] R. Ho, K. Mai, M. Horowitz. The Future of Wires. *Proc. of the IEEE*, pages 490–504, January 2001.

[62] D. Sylvester, K. Keutzer. The Future of Wires. *IEEE Transactions on Computer-Aided Desig of Integrated Circuits and Systems*, pages 242–252, Feb 2000.

[63] D. M. Chapiro. *Globally-Asynchronous LocallySynchronous Systems*. PhD thesis, Stanford University, October 1984.

[64] D. Bertozzi, L. Benini, G. De Micheli. *Network-on-Chip Design for Gigascale Systems-on-Chip*, chapter 21. CRC Press, 2006. ISBN 0-8493-2824-1.

[65] M.T. Rose. *The Open Book - A Practical Perspective on OSI.* 1990.

[66] M. Sgroi, M. Sheets, A. Mihal, K. Keutzer, S. Malik, J.M. Rabaey, A.L. Sangiovanni-Vincentelli. Addressing the System-on-a-Chip Interconnect Woes Through Communication-Based Design. In *Proc. of the Design Automation Conference (DAC)*, 2001.

[67] E. Rijpkema, K.G.W. Goossens, A. R¢adulescu, J. Dielissen, J. van Meerbergen, P. Wielage, and E. Waterlander. Trade Offs in the Design of a Router with Both Guaranteed and Best-Effort Services for Networks on Chip. In *Proc. Int. Conf. on Design, Automation and Test in Europe (DATE)*, 2003.

[68] H. Meyr. *Digital Receivers*, 1998.

[69] W.J. Dally and B. Towles. Route Packtes, Not Wires: ON-Chip Interconnection Networks. In *Proc. of the Design Automation Conference (DAC)*, 2001.

[70] J.A. Rowson and A. Sangiovanni-Vincentelli. Interface-Based Design. In *Proc. of the Design Automation Conference (DAC)*, 1997.

[71] A.S. Tannebaum. *Computer Networks*. Printice-Hall. Inc, 1994.

[72] A. Varma, C.S. Raghavendra. Interconnection Networks for Multiprocessors and Multicomputers: Theory and Practice. Technical report, 1994.

[73] G. Nong, M. Hamdi. On the provision of quality-of-service guarantees for input queued switches. *IEEE Communications Magazine*, 38(12):62–69, December 2000.

[74] Y. Tamir, G. Frazier. High performance multi-queue buffers for VLSI communication Switches. In *Proc. Of 15th Ann. Symp. On Comp. Arch.*, 1988.

[75] S. Kumar, A. Jantsch, J.-P. Soininen, M. Forsell, M. Millberg, J. Öberg, K. Tiensyrjä, A. Hemani. A Network on a Chip Architecture and Design Methodology. In *Proc. IEEE Computer Society Annual Symposium on VLSI*, April 2002.

[76] J.A.J. Leijten, J.L. Van Meerbergen, A.H. Timmer, J.A.G. Jess. Prophid: A Platform-Based Design Method. *Design Automation of Embedded Systems*, 6(1):5–37, 2000.

[77] P. Guerrier, A. Greiner. A Generic Architecture for On-Chip Packet-Switched Interconnections. In *Proc. Int. Conf. on Design, Automation and Test in Europe (DATE)*, 2000.

[78] F. Karim, A. Nguyen, S. Dey, R. Rao. On-Chip Communication Architecture for OC-768 Network Processors. In *Proc. of the Design Automation Conference (DAC)*, 2001.

[79] STNoC: Building a New System-on-Chip Paradigm. White Paper, Dec 2005.

[80] A. Ferrari and A. Sangiovanni-Vincentelli. System design: Traditional concepts and new paradigms. In *Proc. IEEE Int. Conference on Computer Design (ICCD)*, pages 2–13, 1999.

[81] A Sangiovanni-Vincentelli. Defining platform-based design. EEDesign, Feb 2002.

[82] L.P. Carloni, F. De Bernardinis, C. Pinello, A.L. Sangiovanni-Vincentelli, M. Sgroi. *Platform-Based Design for Embedded Systems*, chapter 22. CRC Press, 2006. ISBN 0-8493-2824-1.

[83] P. Magarshack, P. Paulin. System-on-chip Beyond the Nanometer Wall. In *Proc. of the Design Automation Conference (DAC)*, 2003.

[84] A. Jantsch, S. Kumar, A. Hemani. The Rugby model: A conceptual frame for the study of modelling, analysis and synthesis concepts of electronic systems. In *Proc. Int. Conf. on Design, Automation and Test in Europe (DATE)*, 1999.

[85] S.A. Edwards, L. Lavagno, E.A. Lee, and A. Sangiovanni-Vincentelli. Design of embedded systems: Formal Models, Validation, and Synthesis. *Proc. of the IEEE*, 85(3):366–390, March 1997.

[86] G. Berry. The Foundations of Esterel. In R. Milner, G. Plotkin, C. Stirling, M. Tofte, editor, *Proof, Language and Interaction: Essays in Honour*. MIT Press, 1998.

[87] Nicolas Halbwachs. *Synchronous Prgramming of Reactive Systems*. Kluwer Academic Publishers, 1993.

[88] A. Benveniste, G. Berry. The Synchronous Approach to Reactive and Real-Time Systems. *Proc. of the IEEE*, 79(9), Sept. 1991.

[89] F. Boussinot and R. de Simone. The esterel language. *Proceedings of the IEEE*, 79(9):1293–1304, September 1991.

[90] Gilles Kahn. The Semantics of a Simple Language for Parallel Programming. In *Information Processing 74*. North-Holland Publishing Company, 1974.

[91] E.A. Lee, D.G. Messerschmitt. Synchronous data fbw. *Proc. of the IEEE*, September 1987.

[92] J.T. Buck. *Scheduling dynamic Dataflow Graphs with Bounded Memory using the Token Flow Model*. PhD thesis, EECS UC Berkeley, 1993.

[93] T. Grötker, R. Schoenen, and H. Meyr. PCC: A Modeling Technique for Mixed Control/Data FLow Systems. *Proceedings of the European Design and Test Conference*, 1997.

[94] C.A.R. Hoare. *Communicating Sequential Processes*. Prentice-Hall International, 1985.

[95] R. Milner. *Communication and Concurrency*. Prentice-Hall International, 1989.

[96] Telecommunication Standardization Sector of ITU. Specification and description language (sdl). ITU–T Recommendation Z.100, International Telecommunication Union, March 1993.

[97] Telecommunication Standardization Sector of ITU. Annex f.3 to recommendation z.100: Sdl formal definition, dynamic semantics. ITU–T Recommendation Z.100, International Telecommunication Union, November 1988.

[98] C. Hewitt. Viewing control structures as patterns of passing messages. Technical report, 1977.

[99] E. Lee, S. Neuendorffer, M. Wirthlin. Actor-oriented Design of Embedded Hardware and Software Systems, June 2003.

[100] J.T. Buck, S. Ha, E.A. Lee, and D.G. Messerschmitt. Ptolemy: A Framework for Simulating and Prototyping Heterogeneous Systems. *Int. Journal of Computer Simulation*, 4:155–182, April 1994.

[101] F. Balarin, Y. Watanabe, H. Hsieh, L. Lavagno, C. Passerone, A. Sangiovanni-Vincentelli. Metropolis: An Integrated Electronic System Design Environment. *IEEE Computer*, 36(4):45–52, April 2003.

[102] D. Gajski, J. Zhu, R. Dömer, A.Gerstlauer, S. Zhao. *SpecC: Specification Language and Methodology*. Kluwer Academic Publishers, 2000.

[103] Open SystemC Initiative. *Functional Specification for SystemC 2.0*, second edition, April 2002.

[104] D. Gajski L. Cai, S. Verma. Comparison of specc and systemc languages for system design.

[105] S. Swan. An introduction to system level modeling in systemc 2.0, May 2001.

[106] The Future of System Level Design: Can We Find the Right Solution at the Right Time? Proc. of the IEEE/ACM/IFIP Int. Conference on Hardware/Software Codesign and System Synthesis, 2003.

[107] R. Zurawski, editor. *Embedded Systems Handbock*. CRC Press, 2006. ISBN 0-8493-2824-1.

[108] D. Becker, R.K. Singh, and S.G. Tell. An Engineering Environment for Hardware/Software Co-Simulation. In *Proc. of the Design Automation Conference (DAC)*, pages 129–134, 1992.

[109] Seamless. *Mentor Graphics, http://www.mentor.com/.*

[110] J. Notbauer, T. Albrecht, G. Niedrist, S. Rohringer. Verification and management of a multimillion-gate embedded core design. In *Proc. of the Design Automation Conference (DAC)*, 1999.

[111] VSI Alliace. *Hardware-dependent Software (HdS) Taxonomy*, version 1.0 edition.

[112] K. Hines, G. Borriello. Dynamic communication models in embedded system co-simulation. In *Proc. of the Design Automation Conference (DAC)*, 1997.

[113] V. Zivojnovic and H. Meyr. Compiled HW/SW Co-Simulation. In *Proc. of the Design Automation Conference (DAC)*, 1996.

[114] A. Hoffmann, T. Kogel, H. Meyr. A Framework for Fast Hardware-Software Co-simulation. In *Proc. Int. Conf. on Design, Automation and Test in Europe (DATE)*, 2001.

[115] R.K. Gupta, C. Coelho, and G. De Micheli. Synthesis and simulation of digital systems containing interacting hardware and software components. In *Proc. of the Design Automation Conference (DAC)*, 1992.

[116] Giovanni De Micheli Rajesh Gupta, Claudionor N. Coelho. Program Implementation Schemes for Hardware-Software Systems. *IEEE Computer*, pages 48–55, January 1994.

[117] A. Österling, Th. Benner, R. Ernst, D. Herrmann, Th. Scholz, and W. Ye. The COSYMA System. In *Hardware/Software Co-Design: Principles and Practice*. Kluwer Academic Publishers, 1997.

[118] J. Madsen, J. Grode, P. V. Knudsen, M. E. Petersen, A. Haxthausen. LYCOS: The Lyngby cosynthesis system. *Design Automation of Embedded Systems*, 2(2):195–235, 1997.

[119] J.M. Daveau, G.F. Marchioro, T. Ben-Ismail, A.A. Jerraya. *Hardware/Software Co-Design and Co-Verification*, volume 8 of *Current Issues in Electronic Modeling*, chapter COSMOS: An SDL Based Hardware/Software Codesign Environment, pages 59–87. Kluwer Academic Publishers, 1997.

[120] D. Gajski, F. Vahid, S. Narayan, J. Gong. System-level exploration with SpecSyn. In *Proc. of the Design Automation Conference (DAC)*, pages 812–817. ACM Press, 1998.

[121] P. Chou, R. Ortega, and G. Borriello. The chinook hardware/software co-synthesis system. Technical report, 1995.

[122] T. Yen, W. Wolf. Communication Synthesis for Distributed Embedded Systems. In *Proc. of the IEEE Int. Conference on Computer Aided Design*, pages 288–294, November 1995.

[123] M. Gasteiner and M. Glessner. Bus-based communication synthesis on system level. 1996.

[124] J. Daveau, T.B. Ismail, A.A. Jerraya. Synthesis of System-Level Communication by an allocation based approach. In *Proc. Int. Symp. on System Synthesis*, pages 150–155, 1995 September.

[125] P.V. Knudsen and J. Madsen. Integrating Communication Protocol Selection with Partitioning in Hardware/Software Codesign. In *Proc. Int. Symp. on System Synthesis*, 1998.

[126] K. Lahiri, A. Raghunathan, S. Dey. Fast Performance Analysis of Bus-Based System-on-Chip Communication Architectures. In *Proc. of the IEEE Int. Conference on Computer Aided Design*, 1999.

[127] K. Lahiri, A. Raghunathan S. Dey. Performance analysis of systems with multi-channel communication architectures. In *Proc. Int. Conf. VLSI Design*, pages 530–537, 2000.

[128] K. Lahiri, A. Raghunathan, S. Dey. Evaluation of the traffic performance characteristics of system-on-chip communication architectures. In *Proc. Int. Conf. VLSI Design*, pages 29–35, 2001.

[129] A. Jalabert, S. Murali, L. Benini, G. De Micheli. xpipesCompiler: A tool for instantiating application specific Networks on Chip. In *Proc. Int. Conf. on Design, Automation and Test in Europe (DATE)*, 2004.

[130] S. Murali, G. De Micheli. SUNMAP: A Tool for Automatic Topology Selection and Generation for NoCs. In *Proc. of the Design Automation Conference (DAC)*, 2004.

[131] D. Shin, A. Gerstlauer, R. Doemer, D.D. Gajski. Automatic Network Generation for System-on-Chip Communication Design. In *Proc. of the IEEE/ACM/IFIP Int. Conference on Hardware/Software Codesign and System Synthesis*, Sep 2005.

[132] R. Thid, M. Millberg, A. Jantsch. Evaluating NoC Communication Backbones with Simulation. In *Norchip Conference*, 2003.

[133] S.G. Pestana, E. Rijpkema, A. Radulescu, O.P. Gangwahl. Cost-Performance Trade-offs in Networks on Chip: A Simulation-Based Approach . In *Proc. Int. Conf. on Design, Automation and Test in Europe (DATE)*, 2004.

[134] K. Van Rompaey, D. Verkest, I. Bolsens., and H. De Man. CoWare - A Design Environment for Heterogeneous Hardware/Software Systems. In *Proc. of the European Design Automation Conference (EuroDAC)*, 1996.

[135] I. Bolsens, H.J. De Man, B. Lin, K. Van Rompaey, et.al. Hardware/software co-design of digital telecommunication systems. *Proc. of the IEEE*, 85(3):391–418, March 1997.

[136] S. Liao, S. Tjiang, and R. Gupta. An Efficient Implementation of Reactivity for Modeling Hardware in the Scenic Design Environement. In *Proc. of the Design Automation Conference (DAC)*, 1997.

[137] Open SystemC Initiative. *http://www.systemc.org*.

[138] J.P. Robelly, G. Fettweis. Hw/Sw Co-exploration at TLM Level for the Implementation of DSP Algorithms into 2Application Specific DSP's using SystemC and LISA. In *Proc. Int. Workshop on Systems, Architecturs, Modeling and Simulation(SAMOS)*, July 2003.

[139] T. Groetker. Systemc 3.0, 2002.

[140] IEEE. *P1666/D2.1 Standard SystemC Language Reference Manual*, 2005.

[141] SystemC Verification Working Group. *SystemC Verification Standard Specification.* Open SystemC Initiative, Version 1.0e edition, April 2003.

[142] A. Rose, S. Swan, J. Pierce, J.-M. Fernandez. *Transaction Level Modeling in SystemC.* SystemC TLM Working Group.

[143] M. Burton, A. Donlin. Transaction Level Modeling: Above RTL Design and Methodology. www.systemc.org, November 2003.

[144] A. Donlin. Transaction Level Modeling: Flows and Use Models. In *Proc. of the IEEE/ACM/IFIP Int. Conference on Hardware/Software Codesign and System Synthesis*, September 2004.

[145] G. Braun, A. Wieferink, O. Schliebusch, A. Nohl, R. Leupers, H. Meyr. Processor/Memory Co-Exploration on Multiple Abstraction Levels. In *Proc. Int. Conf. on Design, Automation and Test in Europe (DATE)*, Munich, March 2003.

[146] A. Wieferink, T. Kogel, R. Leupers, G. Ascheid, H. Meyr, G. Braun, A. Nohl. A System Level Processor/Communication Co-Exploration Methodology for Multi-Processor System-on-Chip Platforms. In *Proc. Int. Conf. on Design, Automation and Test in Europe (DATE)*, Februry 2004.

[147] F. Ghenassia (Ed.). *Transaction Level Modeling with SystemC.* Springer, 2005.

[148] M. Birnbaum, H. Sachs. How VSIA answers the SOC dilemma. *IEEE Computer*, 32(6):42–50, Jun 1999.

[149] A. Haverinen, M. Leclercq, N. Weyrich, D. Wingard. *White Paper for SystemC based SoC Communication Modeling for the OCP Protocol*, 2003. www.ocpip.org.

[150] T. Kogel, A. Haverinen, J. Aldis. OCP TLM for Architectural Modeling, July 2005. OCPIP whitepaper, www.ocpip.org.

[151] A. Haverinen et al. *A SystemC OCP Transaction Level Communication Channel*, Feb 2006. version 2.1.2, www.ocpip.org.

[152] O. Ogawa, K. Shinohara, Y. Watanabe, H. Niizuma, T. Sasaki, Y. Takai, S. Bayon de Noyer and P. Chauvet. A Practical Approach for Bus Architecture Optimization at Transaction Level. In *Proc. Designers' Forum, Int. Conf. on Design, Automation and Test in Europe (DATE)*, 2003.

[153] T. Kogel, M. Braun. Virtual Prototyping of Embedded Platforms for Wireless and Multimedia. In *DATE*, March 2006. invited paper.

[154] W. Cesario, A. Baghdadi, L. Gauthier, D. Lyonnard, G. Nicolescu, Y. Paviot, S. Yoo, A. Jerraya, M. Diaz-Nava. Component-Based Design Approach for Multicore SoCs. In *Proc. of the Design Automation Conference (DAC)*, 2002.

[155] M.-A. Dziri, W. Cesário, F.R. Wagner, A.A. Jerraya. Unified Component Integration Flow for Multi-Processor SoC Design and Validation. In *Proc. Int. Conf. on Design, Automation and Test in Europe (DATE)*, 2004.

[156] W.O. Cesario, F.R. Wagner, A.A. Jerraya. *Hardware/Software Interface Design for SoC*, chapter 24. CRC Press, 2006. ISBN 0-8493-2824-1.

[157] D. Lyonnard, S. Yoo, A. Baghdadi, A.A. Jerraya. Automatic Generation of Application-Specific Architectures for Heterogeneous Multiplrocessor System-on-Chip. In *Proc. of the Design Automation Conference (DAC)*, 2001.

[158] S. Yoo, G. Nicolescu, D. Lyonnard, A. Baghdadi, A.A. Jerraya. A Generic Wrapper Architecture for Multi-Processor SoC Cosimulation and Design. In *Proc. Int. Symp. on Hardware/Software Codesign (CODES)*, 2001.

[159] S. Yoo, I. Bacivarov, A. Bouchhima, Y. Paviot, A.A. Jerraya. Building Fast and Accurate SW Simulation Models Based on Hardware Abstraction Layer and Simulation Environment Abstraction Layer. In *Proc. Int. Conf. on Design, Automation and Test in Europe (DATE)*, 2003.

[160] A. Jerraya. Application specific multi-processor soc. MPSOC seminar, July 2002. Presentation.

[161] W.O. Cesário, D. Lyonnard, G Nicolescu, Y. Paviot, S. Yoo, and A. Jerraya, L. Gauthier, M. Diaz-Nava. Multiprocessor SoC Platforms: A Component-Based Design Approach. *IEEE Design & Test of Computers*, 19(6):52–63, November-December 2002.

[162] A. Nieuwland, J. Kang, O.P. Gangwal, R. Sethuraman, N. Busa, K. Goossens, R.P. Llopis, P. Lippens. C-HEAP: A Heterogeneous Multi-Processor Architecture Template and Scalable and Flexible Protocol for the Design of Embedded Signal Processing Systems. *Design Automation of Embedded Systems*, 2002.

[163] P. van der Wolf, E. de Kock, T. Henriksson, W. Kruijtzer, G. Essink. Design and Programming of Embedded Multiprocessors: An Interface-Centric Approach. In *Proc. of the IEEE/ACM/IFIP Int. Conference on Hardware/Software Codesign and System Synthesis*, 2004.

[164] P. van der Wolf, E. de Kock, T. Henriksson, W. Kruijtzer, G. Essink. *Design and Programming of Embedded Multiprocessors: An Interface-Centric Approach*, chapter 25. CRC Press, 2006. ISBN 0-8493-2824-1.

[165] V. Reyes, T. Bautista, G. Marrero, A. Nunez, W. Kruijtzer. A Multicask Inter-Task Communication Protocol for Embedded Multiprocessor Systems. In *Proc. of the IEEE/ACM/IFIP Int. Conference on Hardware/Software Codesign and System Synthesis*, September 2005.

[166] A. Pinto, L. Carloni, A. Sangiovanni-Vincentelli. Constraint-Driven Communication Synthesis. In *Proc. of the Design Automation Conference (DAC)*, June 2002.

[167] L.P. Carloni, F. De Bernardinis, Alberto Sangiovanni-Vincentelli, M. Sgroi. The Art and Science of Integrated Systems Design. In *Proceedings of the 28th European Solid-State Circuits Conference*, September 2002.

[168] D. Bertozzi, A. Jalabert, S. Murali, R. Tamhankar, S. Stergiou, L. Benini, G. De Micheli. NoC Synthesis Flow for Customized Domain Specific Multi-Processor Systems-on-Chip. *IEEE Micro*, 22(5):46–55, Sep-Oct 2004.

[169] J.-P. Soininen, H. Heusala. *A Design Methodology for NoC-based Systems*, chapter 2, pages 19–38. Kluwer Academic Publishers, 2003.

[170] M. Gries. Methods for Evaluating and Covering the Design Space during Early Design Development, July 2004.

[171] B. Kienhuis, E. Deprettere, K. Vissers, P. van der Wolf. An Approach for Quantitative Analysis of Application-Specific Dataflow Architectures. In *IEEE International Conference on Application-Specific Systems, Architectures and Processors*, 1997.

[172] Balarin et al. *Hardware-Software Co-Design of Embedded Systems : The Polis Approach.* Kluwer Academic Publishers, 1997.

[173] P. Lieverse, P. and van der Wolf, E. Deprettere, K. Vissers. A Methodology for Architecture Exploration of Heterogeneous Signal Processing Systems. *Journal of VLSI Signal Processing for Signal, Image and Video Technology*, 29(3):197–207, November 2001.

[174] P. Lieverse, P van der Wolf, E. Deprettere,K Vissers. A Methodology for Architecture Exploration of Heterogeneous Signal Processing Systems. In *Proc. IEEE Int. Workshop on SIgnal Processing Systems (SIPS)*, 1997.

[175] E.A. de Kock and W.J.M. Smits and P. van der Wolf and J.-Y. Brunel and W.M. Kruijtzer and P. Lieverse and K.A. Vissers and G. Essink. YAPI: application modeling for signal processing systems. In *Proc. of the Design Automation Conference (DAC)*, pages 402–405. ACM Press, 2000.

[176] R.A. Uhlig, T.N. Mudge. Trace-driven Memory Simulation. *ACM Computing Surveys*, 29(2):128–170, June 1997.

[177] A.D. Pimentel, L.O. Hertzberger, P. Lieverse, P. van der Wolf, E.F. Deprettere. Exploring Embedded-Systems Architectures with Artemis. *IEEE Computer*, 34(11):57–63, November 2001.

[178] A.D. Pimentel, S. Polstra, F. Terpstra, A.W. van Halderen, J.E. Coffland and L.O. Hertzberger. *Embedded Processor Design Challenges: Systems, Architectures, MOdeling, and Simulation (SAMOS)*, chapter Towards Effi cient Design Space Exploration of Heterogeneous Embedded Media Systems, pages 57–73. LNCS, 2002.

[179] A.D. Pimentel, C. Erbas. An IDF based Trace Transformation Method for Communication Refinement. In *Proc. of the Design Automation Conference (DAC)*, June 2003.

[180] C. Erbas, S.C. Erbas, A.D. Pimentel,. A Multiobjective Optimization Model for Exploring Multiprocessor Mappings of Process Networks. In *Proc. of the IEEE/ACM/IFIP Int. Conference on Hardware/Software Codesign and System Synthesis*, October 2003.

[181] V.D. Zivkovic, P. van der Wolf, E.F. Deprettere, E.A. de Kock. Design Space Exploration of Streaming Multiprocessor Architectures. In *Proc. IEEE Int. Workshop on SIgnal Processing Systems (SIPS)*, October 2002.

[182] V.D. Zivkovic, E.F. Deprettere, E.A. de Kock, P. van der Wolf. Fast and Accurate Multiprocessor Architecture Exploration with Symbolic Programs. In *Proc. Int. Conf. on Design, Automation and Test in Europe (DATE)*, 2003.

[183] L. Thiele, E. Wandeler. *Performance Analysis of Distributed Embedded Systems*, chapter 15. CRC Press, 2006. ISBN 0-8493-2824-1.

[184] M. Gries, C. Kulkarni, C. Sauer, K. Keutzer. Comparing Analytical Modeling with Simulation for Network Processors: A Case Study. In *Proc. Int. Conf. on Design, Automation and Test in Europe (DATE)*, March 2003.

[185] VCC: Virtual Component Co-Design. *Cadence, http://www.cadence.com.*

[186] B.D. Bowen. Felix to Move Codesign from Problem to Solution. *Cadence Plugged-In Magazine*, 3(1), April 1998.

[187] J.R. Bammi, E. Harcourt, W. Kruijtzer, L. Lavagno, M.T. Lazarescu. Software Performance Estimation Strategies in a System-Level Design Tool. In *Proc. Int. Symp. on Hardware/Software Codesign (CODES)*, 2000.

[188] J.-Y. Brunel and W.M. Kruijtzer and H.J.H.N. Kenter and F. Pétrot and L. Pasquier and E.A. de Kock and W.J.M. Smits. COSY communication IP's. In *Proc. of the Design Automation Conference (DAC)*, pages 406–409. ACM Press, 2000.

[189] P. Kajfasz, M. Bourdelles. SYNTEL: A Synchronous Co-design Environment for the Synthesis of Wireless Telecommunication Protocols. In *Proc. Int. Workshop on Systems, Architecturs, Modeling and Simulation(SAMOS)*, pages 135–141, 2003.

[190] F. Balarin, L. Lavagno, C. Passerone, A. Sangiovanni-Vincentelli, M. Sgroi, Y. Watanabe. Modeling and designing heterogeneous systems. In J. Cortadella, A. Yakovlevm G, Rozenberg, editor, *Concurrency and Hardware Design*, Lecture Notes in Computer Science, pages 228–273. Springer, 2002.

[191] G. Goessler, A. Sangiovanni-Vincentelli. Compositional Modeling in Metropolis. In *Proc. EMSOFT'02*, October 2002.

[192] F. Balarin, L. Lavagno, C. Passerone, Y. Watanabe. Processes, interfaces and platforms. Embedded software modeling in Metropolis. In *Proc. EMSOFT'02*, October 2002.

[193] W. Mueller, R. Dömer, A. Gerstlauer. The Formal Execution Semantics of SpecC. In *Proc. Int. Symp. on System Synthesis*, 2002.

[194] R. Dömer. *System-level Modeling and Design with the SpecC Language*. PhD thesis, University Dortmund, 2000.

[195] A. Gerstlauer, H. Yu, D.D. Gajski. RTOS Modeling for System Level Design. In *Proc. Int. Conf. on Design, Automation and Test in Europe (DATE)*, 2003.

[196] H. Yu, A. Gerstlauer, D. Gajski. RTOS Scheduling in Transaction Level Models. In *Proc. of the IEEE/ACM/IFIP Int. Conference on Hardware/Software Codesign and System Synthesis*, 2003.

[197] J.M. Paul, and D.E. Thomas. A Layered, Codesign Virtual Machine Approach to Modeling Computer Systems. In *Proc. Int. Conf. on Design, Automation and Test in Europe (DATE)*, 2002.

[198] E. Donald, J.M. Thomas, Paul S.N. Peffers. Frequency interleaving as a codesign scheduling paradigm. In *Proc. Int. Symp. on Hardware/Software Codesign (CODES)*, 2000.

[199] M. JoAnn S. Paul Andrew Cassidy and Donald E. Thomas. Layered, multi-threaded, high-level performance design. In *Proc. Int. Conf. on Design, Automation and Test in Europe (DATE)*, 2003.

[200] J.M. Paul, A. Bobrek, J.E. Nelson, J.J. Pieper, D.E. Thomas. Schedulers as Model-Based Design Elements in Programmable Heterogeneous Multiprocessors. In *Proc. of the Design Automation Conference (DAC)*, 2003.

[201] A. Cassidy. High-Level Performance Modeling and Design Exploration. Technical report, Electrical and Computer Engineering Department, Carnegie Mellon University, 2002.

[202] J.M. Paul, D.E. Thomas, A. Bobrek. Benchmark-Based Design Strategies for Single-Chip Heterogeneous Multiprocessors. In *CODES+ISSS*, September 2005.

[203] D. Quinn, B. Lavigueur, G. Bois, M. Aboulhamid. A System Level Exploration Platform and Methodology for Network Applications Based on Configurable Processors. In *Proc. Int. Conf. on Design, Automation and Test in Europe (DATE)*, 2004.

[204] P.G. Paulin, C. Pilkington, M. Langevin, E. Bensoudane, G. Nicolescu. Parallel Programming Models for a Multi-Processor SoC Platform Applied to High Speed Traffic Management. In *Proc. of the IEEE/ACM/IFIP Int. Conference on Hardware/Software Codesign and System Synthesis*, 2004.

[205] P.G. Paulin, C. Pilkington, M. Langevin, E. Bensoudane, D. Lyonnard, G. Nicolescu. *A Multiprocessor SoC Platform and Tools for Communications Applications*, chapter 26. CRC Press, 2006. ISBN 0-8493-2824-1.

[206] M. Coppola, S. Curaba, M.D. Grammatikakis, G. Maruccia, F. Papariello. OCCN: A Network-On-Chip Modeling and Simulation Framework. In *Proc. Int. Conf. on Design, Automation and Test in Europe (DATE)*, 2004.

[207] M. Coppola, S. Curaba, M.D. Grammatikakis, G. Maruccia, F. Papariello. The OCCN user manual. Technical report.

[208] M. Coppola, S. Curaba, M.D. Grammatikakis, G. Maruccia. IPSIM: SystemC 3.0 enhancements for communication refinement. In *Proc. Int. Conf. on Design, Automation and Test in Europe (DATE)*, 2003.

[209] W. Klingauf, R. Guenzel, O. Bringmann, P. Parfuntseu, M. Burton. Greenbus, 2006. www.greensocs.com.

[210] MPArm. *http://www-micrel.deis.unibo.it/sitonew/projects/mparm.html.*

[211] S. Mahadevan, M. Storgaard, J. Madsen, and K. M. Virk. ARTS: A system-level framework for modeling mpsoc components and analysis of their causality. In *13th IEEE International Symposium on Modeling, Analysis, and Simulation of Computer and Telecommun icationSystems (MASCOTS)*. IEEE Computer Society, sep 2005.

[212] J. Madsen, K. Virk, M.J. Gonzalez. Abstract RTOS Modelling for Multiprocessor System-on-Chip. In *International Symposium on System-on-Chip*, pages 147–150. IEEE, nov 2003.

[213] J. Madsen, S. Mahadevan, K. Virk, M.J. Gonzalez. Network-on-Chip Modeling for System-Level Multiprocessor Simulation. In *The 24th IEEE International Real-Time Systems Symposium*, pages 265–274. IEEE Computer Society, Dec 2003.

[214] K. Virk, S. Mahadevan, J. Madsen. Abstract System-on-Chip Modelling in SystemC, February 2004. European SystemC User Group Meeting, www-ti.informatik.uni-tuebingen.de/ systemc/systemc.html.

[215] G. Post, A. Müller, R. Schoenen. Object-Oriented Design of ATM Switch Hardware in a Telecommunication Network Simulation Environment. In *Proc. Int. Symp. on Hardware/Software Codesign (CODES)*, 1998.

[216] G. Post. *Methodik zur objektorientierten Modellierung und Hardware/Software–Covalidation komplexer Telekommunikationssysteme*. PhD thesis, RWTH Aachen, 1999. ISBN 3-8265-6555-X.

[217] A. Kroll. *Verifikationseffiziente Implementierung von Verkehrsmanagement Funktionalität in ATM-Vermittlungsstellen*. PhD thesis, RWTH Aachen, 2001. ISBN 3-8265-8849-5.

[218] T. Kogel, A. Wieferink, H. Meyr, A. Kroll. SystemC based Architecture Exploration of a 3D Graphic Processor. In *Proc. IEEE Int. Workshop on SIgnal Processing Systems (SIPS)*, September 2001.

[219] D. Bussaglia T. Kogel. Systemc based design of an ip forwarding chip with cocentric system studio. In *Synopsys User Group Europe (SNUG)*, March 2002.

[220] OPNET. *http://www.opnet.com.*

[221] A. Hofmann, H. Meyr, R. Leupers. *Architecture Exploration for Embedded Processors with LISA*. Kluwer Academic Publishers, 2002. ISBN 1-4020-7338-0.

[222] O. Schliebusch, A. Chattopadhyay, M. Steinert, G. Braun, A. Nohl, R. Leupers, G. Ascheid, H. Meyr. RTL Processor Synthesis for Architecture Exploration and Implementation. In *Proc. Designers' Forum, Int. Conf. on Design, Automation and Test in Europe (DATE)*, Feburary 2004.

[223] T. Kempf. System Level Design of an Optimized Network-on-Chip Architecture for an IPv4 DiffServ Platform. Diploma Thesis, December 2003.

[224] A. Wieferink, M. Doerper, R. Leupers, Gerd Ascheid, H. Meyr, T. Kogel. Early ISS integration into Network-on-Chip Designs. In *Proc. Int. Workshop on Systems, Architecturs, Modeling and Simulation(SAMOS)*, July 2004.

[225] A. Jantsch. *Models of Embedded Computation*, chapter 4. CRC Press, 2006. ISBN 0-8493-2824-1.

[226] I.N. Bronstein, K.A. Semendjajev. *Taschenbuch der Mathematik*. Verlag Harri Deutsch, 1998.

[227] IEEE Standard VHDL Language Reference Manual. IEEE Std 1076, March 1987.

[228] W. Mueller, J. Ruf, D. Hoffmann, J. Gerlach, T. Kropf, W. Rosenstiehl. The Simulation Semantics of SystemC. In *Proc. Int. Conf. on Design, Automation and Test in Europe (DATE)*, 2001.

[229] D.W. Jones. An Empirical Comparison of Priority Queue Algorithms, April 1986.

[230] R. Brown. Calendar queues: A fast O(1) priority queue implementation for the simulation event set problem. *Communications of the ACM*, 31(10):1220 – 1227, Oct 1988.

[231] K. Chung, J. Sang, V. Rego. A Performance Comparison of Event Calendar Algorithms: an Empirical Approach. *Software - Practice and Experience*, 23(10):1107–1138, 1993.

[232] N. Weyrich, A. Haverinen. *A SystemC Generic Transaction Level Communication Channel*, 2003. www.ocpip.org.

[233] B. Vanthournout, S. Goossens, T. Kogel. Developing Transaction-level Models in SystemC. White Paper, CoWare Inc., June 2004.

[234] G. Post R. Schoenen. Static bandwidth allocation for input-queued switches with strict qos bounds. In *IEEE Broadband Switching Systems*, 1999.

[235] T. Anderson, S. Owicki, J. Saxe, and C. Thacker. High-speed switch scheduling for localarea networks, November 1993.

[236] N. McKeown. *Scheduling Algorithms for Input-Queued Cell Switches*. PhD thesis, EECS UC Berkeley, 1995.

[237] Balaji Prabhakar, Nick McKeown, and Ritesh Ahuja. Multicast scheduling for input-queued switches. *IEEE Journal on Selected Areas in Communications*, 15(5):855–866, 1997.

[238] R. Schoenen, G. Post, and G. Sander. Prioritized arbitration for inputqueued switches with 100% throughput, 1999.

[239] R. Callon, A. Viswanathan, E. Rosen. Multiprotocol Label Switching Architecture. RFC 3031, July 1998.

[240] Tim Bray, Jean Paoli, C.M. Sperberg-McQueen, and Eve Maler. Extensible markup language (xml) 1.0 (second edition) w3c recommendation, 6 October 2000.

[241] M. Doerper. Development of a SystemC based Modular Simulation Framework for System Level Exploration of Network-on-Chip Architectures. Diploma Thesis, September 2003.

[242] Telecommunication Standardization Sector of ITU. Message Sequence Chart (MSC). ITU–T Recommendation Z.120, International Telecommunication Union, March 1993.

[243] TAU SDL suite. *Telelogic, http://www.telelogic.com.*

[244] T. Kogel, M. Doerper, T. Philipp, O. Zerres. *GRACE++ User Guide.* ISS, RWTH Aachen.

[245] GTKWave. *http://www.cs.man.ac.uk/apt/tools/gtkwave/.*

[246] gnuplot. *http://www.gnuplot.info.*

[247] graphviz. *http://www.research.att.com/sw/tools/graphviz.*

[248] J.P.C. Kleijnen. Sensitivity Analysis and Optimization in Simulation: Design of Experiments and Case Studies. In *Proc. of the Winter Simulation Conference*, 1995.

[249] An Architecture for Differentiated Services. *http://www.ietf.org/rfc/rfc2475.txt.*

[250] Requirements for IP Version 4 Routers. *http://www.ietf.org/rfc/rfc1812.txt.*

[251] J. Heinanen and R. Guerin. *A Single Rate Three Color Marker.* Telia Finland, University of Pennsylvania, September 1999. RFC 2697.

[252] S. Floyd, and V. Jacobson. Random early detection gateways for congestion avoidance. *IEEE/ACM Transactions on Networking*, 1(4):379–413, August 1993.

[253] R. Nennen. A SystemC based. Diploma Thesis, Feburary 2003.

[254] M. Gries. *Algorithm-Architecture Trade-offs in Network Processor Design.* PhD thesis, Swiss Federal Institute of Technology Zurich, May 2001.

[255] The Network Processor Forum. *founded by CSIX/CPIX members in 2001 http://www.npforum.org.*

[256] Intel Network Processors. *http://developer.intel.com/design/network/products/npfamily/.*

[257] A. Jantsch, H. Tenhunen. *Will Networks on Chip Close the Productivity Gap?*, chapter 1, pages 3 – 18. Kluwer Academic Publishers, 2003.

[258] M. Doerper. A SystemC based Stochastical Simulation Environment for System Level Simulations. Student Thesis, Feburary 2003.

[259] H. Meyr. Keynote Speech : System-on-chip for communications : The dawn of ASIPs and the dusk of ASICs. Keynote of IEEE Int. Workshop on SIgnal Processing Systems (SIPS), August 2003.

[260] B. Bailey. Property Based Verification for SoC. Int. Symp. on System-on-Chip (SoC), November 2003. Invited Talk.

About the Authors

Tim Kogel received his Dipl. Ing. degree in Electrical Engineering from Aachen University of Technology (RWTH), Aachen, Germany, in 1999. During his time as a Ph.D. student at the same university he has authored numerous publications on System Level Design of Multi-Processor System-on-Chip platforms. In 2005 he received is PhD degree from RWTH Aachen with honors.

Today, he is a Solution Specialist working in the product engineering team at CoWare Inc. In this position he represents CoWare in the SystemC Transaction Level Modeling related standardization committees from OCPIP and OSCI as well as in the technical program committees of ISSS/CODES and DATE.

Contact Information:
Tim Kogel
CoWare, Inc.
Dennewartstrasse 25-27
52068 Aachen, Germany
email: tim.kogel@CoWare.com
web: http://www.CoWare.com

Heinrich Meyr received his M.Sc. and Ph.D. from ETH Zurich, Switzerland. He spent over 12 years in various research and management positions in industry before accepting a professorship in Electrical Engineering at Aachen University of Technology (RWTH Aachen) in 1977. He has worked extensively in the areas of communication theory, digital signal processing and CAD tools for system level design for the last thirty years. His research has been applied to the design of many industrial products. At RWTH Aachen he is a co-director of the institute for integrated signal processing system (ISS) involved in the analysis and design of complex signal processing systems for communication applications.

He was a co-founder of CADIS GmbH (acquired 1993 by Synopsys, Mountain View, California) a company which commercialized the tool suite COSSAP. In 2001 he has co-founded LISATek Inc., a company with breakthrough technology to design application specific processors. In February 2003 LISATek has been acquired by CoWare, an acknowledged leader in the area of system level design. At CoWare Dr. Meyr has accepted the position of Chief Scientist.

Dr. Meyr has published numerous IEEE papers and holds many patents. He is author (together with Dr. G. Ascheid) of the book S̈ynchronization in Digital Communications;· Wiley 1990 and of the book D̈igital Communication Receivers. Synchronization, Channel Estimation, and Signal Processing(̈together with Dr. M. Moeneclaey and Dr. S. Fechtel), Wiley, October 1997. He has received two IEEE best paper awards.

In 1998 he was a visiting scholar at UC Berkeley's wireless research center (BWRC). He was elected as the M̈c Kay distinguished lecturerät the EE department of the UC Berkeley for the spring term 2000. Dr.Meyr is also the recipient of the prestigious V̈odafone Innovation Prizeför the year 2000. The Vodafone prize is awarded for outstanding contribution to the area of wireless communication .

As well as being a Fellow of the IEEE he has served as Vice President for International Affairs of the IEEE Communications Society

Rainer Leupers received the Diploma and Ph.D. degrees in Computer Science with honors from the University of Dortmund, Germany, in 1992 and 1997. From 1997-2001 he was a senior research engineer at the Embedded Systems group at the University of Dortmund. Between 1999-2 001 he was also a project manager at ICD, where he headed the development of custom C compilers and other industrial software tool projects. In 2002, Dr. Leupers joined RWTH Aachen University as a professor for Software for Systems on Silicon. His research and teaching activities revolve around software development tools, processor architectures, and electronic design automation for embedded systems, with emphasis on C compilers for application specific processors in the areas of signal processing and networking. He authored several books and numerous technical papers on software tools for embedded processors, and he served in the program committees of leading EDA an d compiler conferences, including DAC, DATE, and ICCAD. Dr. Leupers received several scientific awards, including Best Paper Awards at DATE 2000 and DAC 2002. He has been a co-founder of LISATek, an EDA tool provider for embedded processor design (acquired by CoWare Inc. in 2003).

Index